# CookLike

## 跟**大廚**做西菜

鄭雅正編著　萬里機構‧飲食天地出版社出版

## AChef

# 目錄 Contents

**004** 前言 Foreword

**第一課　西廚入門 Start Cooking**

**008** I. 廚房分工與權責
Division of Labor

**008** I.1 西廚職位分工結構表
Classic Kitchen Organization
Chart

**009** I.2 廚房員工的權責
Job Description

**010** II. 餐單設計 Menu Designing

**010** II.1 設計餐單的考慮因素
Factors Affecting Menu Design

**012** II.2 餐牌之種類 Types of Menu

**014** III. 西餐的基本烹調方法
Basic Cooking Methods of
Western Cuisine

**第二課　廚房裏的秘密 Cooking Basics**

**018** I. 工具概覽
Cooking Tools Galleries

**020** I.1 廚刀分類
Classification of Knives

**021** I.2 基本刀章法
Basic Cutting Skills

**022** II. 材料的選購與烹調關係
Good Ingredients, Better

**023** III. 葡萄酒作為烹煮用途
Cook with Grape Wine

**024** IV. 食材概覽
Grains, Pasta & Vegetables

**024** IV.1 穀物與麵概覽
Grains and Pasta Galleries

**025** IV.2 蔬菜概覽
Vegetables Galleries

**026** V. 香草與香料的運用
The Use of Herbs & Spices

**026** V.1 西餐的基本烹調方法
Basic Cooking Methods of
Western Cuisine

**028** VI. 基本湯底 Basic Stocks

**028** a. 香草束 Bouquet Garni

**028** b. 白湯底 White Stock

**029** c. 黃湯底 Brown Stock

**030** d. 魚湯底 Fish Stock

**031** e. 湯底的貯藏 Storage of Stock

**031** f. 油麵糊 / 麵撈 Types of Roux

**032** VII. 醬汁與調味
Precious Sauces and Dressings

**032** VII.1 不同醬汁的烹煮法
Assorted Sauces

**032** a 法式奶白汁 Béhamel Sauce

**033** b 高湯白汁 Velouté Sauce

**034** c 基本黃汁
Basic Brown Sauce

**035** VII.2 不同的調味醬
Assorted Dressing

**035** a 牛油醬汁 - 荷蘭汁
Butter Sauce-Basic Hollandaise

**036** b 蛋黃醬汁 Mayonnaise

**037** c 油醋汁 Vinaigrette

**038** VIII. 麵糰與麵條的製作
Making Pasta Dough, Rolling
and Cutting Pasta

**038** a 麵糰製作
Making Pasta Dough

**039** b 麵條製作
Rolling and Cutting Pasta

**040** IX. 馬鈴薯蓉與中東粗麥粉
Mashed Potatoes & Cous

**040** a 馬鈴薯蓉 Mashed Potatoes

**041** b 粗麥粉 Cous cous

**第三課　繁忙的廚房料理製作
Cooking Workshops**

**I. 湯 Delicious Soups**

**044** 牛肉清湯 Beef Consomm

**046** 法國洋葱湯 French Onion Soup

**048** 甘筍茸湯 Puréed Carrot Soup

**050** 海鮮大燴配蒜頭甜椒醬
Creole Bouillabaisse with Rouille
Sauce

**052** 英式海鮮周打湯
England Seafood Chowder

## II. 沙律及蔬菜類
## Refreshing Salads and Vegetables

054 凱撒沙律 Caesar Salad

056 牛油果香橙沙律
Avocado and Orange Salad

058 新式吞拿魚沙律
Modern Salad Nicoise

060 鳳尾蝦喀嗲 Shrimp Cocktail

062 法式燴菜 Zesty Ratatouille

064 菠菜醬糕 Spinach Timbale

## III. 飯，穀物和麵
## Rice, Grains and Pasta

066 腰果香飯 Cashew-rice Pilaf

068 白菌意大利飯 Mushroom Risotto

070 西班牙海鮮飯
Spanish Paella with Chicken,Prawn and Squid

072 玉米糕 Polenta

074 肉醬意大利粉
Spaghetti Bolognese

076 香草醬扁麵
Linguine with Pesto Sauce

078 煙肉蛋黃汁螺絲粉
Fusilli Carbonara

## IV. 魚貝海鮮 Fish and Seafood

080 香草醃三文魚
Gravadlax with Mustard Dill Sauce

082 清蒸魚辮 Steaming Fish Plaits

084 炭烤全魚 Barbecuing Whole Fish

086 紙包焗魚 Baking en Papillote

088 脆炸魚柳
Deep-frying Fish in Batter

090 美式煎三文魚
Broiling Cajun-style Salmon Steak

092 龍蝦 "米多" Lobster Thermidor

094 香草蒜茸牛油焗青口
Mussels with Garlic Herb Butter

## V. 家禽野味 Poultry and Game

096 烤全雞
Roasting a Whole Chicken

098 釀雞卷 Stuffed Chicken Roll

100 紅酒燴雞 Coq au Vin

102 雞肝醬 Chicken Liver Pâte

104 鑊煎鴨胸 Pan-fried Duck Breast

106 野米鵝肝蘑菇釀鵪鶉
Boned Quail, Filled with Wild Rice, Foie Gras and Mushroom

## VI. 肉類
## Meat (Beef, Veal, Lamb and Pork)

108 韃靼牛肉 Beef Steak Tartare

110 烤牛柳黯
Roasting Beef Tenderloin

112 焗威靈頓牛柳
Baked Beef Wellington

114 意式煎小牛肉 Saltimbocca

116 燴牛仔膝 Osso Buco

118 皇冠焗羊排 Crown Roast of Lamb

120 釀烤羊髀
Stuffed and Roasting Leg of Lamb

122 牛奶燴豬排 Braising Pork in Milk

## VII. 甜品類 Desserts

124 紅酒蜜桃 Peach in Red Wine

126 雜莓糖霜蛋白籃
Berries on Meringue Nest

128 法式燉蛋 Crème Brûlée

130 朱古力慕士
Brown Chocolate Mousse

132 焗梳芙厘 Hot Soufflé

134 法式水果撻 Tarte Tatin

136 西柚碎雪
Iced Coulis of Grapefruit-making Granita

138 脆米紙焗釀蘋果 Apple Croustade

140 凍忌廉芝士餅
Chilled Cream Cheese Cake

142 常用術語 Common Terminology

143 換算表 Convention Table

144 鳴謝 Acknowledgements

# 前言

我從小便愛烹調料理，只是多在家中幫忙，不過從幫忙的過程中，卻給了我很多基本烹調概念及知識，奠定了日後在學習料理上的良好基礎。

回顧過去23年的烹調歲月，主要從事於西式料理的行業，期間也會處理一些東南亞料理，從中得以窺見各種地方菜式的不同風格及烹調技巧，因而融入到自身做菜特色之中。

在學徒時代，學習的過程比較辛苦，不過現在回憶起來，仍然令我十分懷念⋯第一次調理湯汁，第一次燴肉；印象難忘，有無數的試練失敗，也有成功的喜悅，回味再三。事實上，從前的師傅廚師大都不會詳細地教你如何烹調，往往只會責罵不斷，為了學有所成，我只好偷偷從旁學習，由於沒有得到詳細解釋及仔細教導，只能"知其形而不知其髓"，所以經常招致失敗。不過把失敗的經驗累積下來的結果，卻構成了個人的烹調風格。

製作西式料理並不太困難，它會有既定的基礎原理：材料與工具認識；基本湯底和醬汁；餐單編排和食品製作，當然份量換算也是不容忽視的。所以必須徹底的學會基本烹調原理作基礎，才可把菜做好。但是沒有把基本菜式弄好，又心多多地學做其他菜式，只會弄巧反拙，這樣絕不能使你的手藝有所進步；相反地，假如你能精通基本的烹調原理，日後只要稍加融匯貫通，便能夠製作出變化多端的佳餚。我的忠告：不要只顧學習新奇花招，必須把基本功學得徹底才是最重要。

謹此希望結集過往烹調經驗，直接告訴大家在西式料理上的基本技巧及秘訣，藉此幫助各位讀者瞭解和認識西式料理。

鄭雅正

# Foreword

Cooking has been my interest since I was a child. Mostly I cooked at home during my childhood. I learned cooking concepts and knowledge from the process of cooking. These concepts and knowledge gave me a good foundation of cooking.

Looking back to the past 23 years of my chef's days, I have been mainly devoting myself to the western cooking. Sometimes I will cook the South East Asian cuisines. In order to get chances of handling cross-countries cuisines, I found different countries had their own cooking style and methods. These findings benefit me to incite my cooking ideas and promote my cooking skills.

Recalling on my apprentice's periods, the learning process was hard but it was over. I miss it very much, especially at the first time I cooked soup and stewed meat. In fact, some cooking experiences were successful but some were not. No matter how good or bad they were, those experiences had deep impressions on my memory. I enjoyed very much. Indeed, chefs in old days did not like teaching the apprentices how to cook the food in detail, in addition, they often scolded young cooks only. Those aggressive apprentices or young cooks who wanted to become chefs at the future would secretly learn from watching besides the chefs. Although the cooks could learn the cooking skills by watching, they did not completely understand. The cooking results were frequently failures because of lack of detailed explanations and delicated teaching from the chefs. Accumulating a series of failing results, I learn more and more that make me build up my cooking style.

It is not difficult to cook the western cuisines if you have a good knowledge of their foundation concepts. They are knowledges of ingredients and tools, basic soups and sauces, menu design and food process. You should also pay much attention to convertible measurements. Apparently, you should learn cooking theories thoroughly as a foundation if you want to cook perfect dishes. If you cannot manage your basic dishes very well, your attempt to cook another countries* cuisine will not be successful. Your cooking skills will not improve and you will not be a professional. On the contrary, if you are familiar with the principles of cooking, you can generate more cooking ideas on handling cuisines smartly. You can also create or change the dishes by your own idea. For my advice, "do not focus on changing dishes only. It is more important to learn the foundation of cooking skills than focus on dishes variations."

This book collects my past cooking experiences. It directly states how to cook the western cuisines by mastering principles and tips of cooking. The aim is to let the readers understand more about the western cooking.

*Julian Cheng*

第一課

西廚入門 *Start Cooking*

# I. 廚房分工與權責
## Division of Labor

西式廚房的分工相當仔細，現介紹一個標準的西廚分工結構，讓年青廚師有個初步認知的概念。

Labor of division in western kitchen is detailed. In order to let young cooks have a brief concept of a classic kitchen, its structure is stated as below:

## I.1 西廚職位分工結構表
### Classic Kitchen Organization Chart

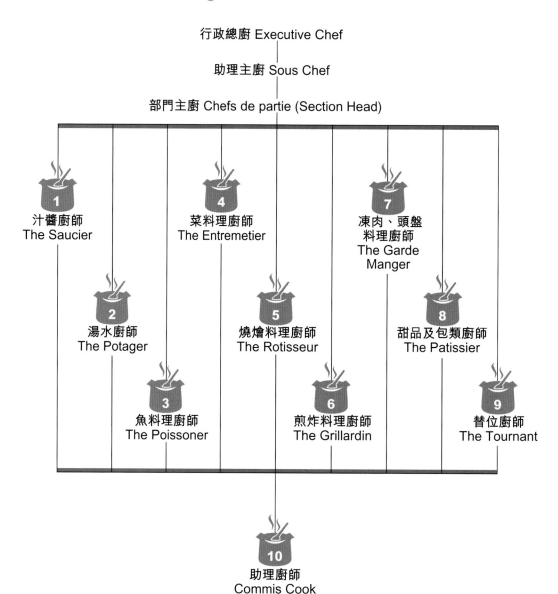

行政總廚 Executive Chef

助理主廚 Sous Chef

部門主廚 Chefs de partie (Section Head)

1 汁醬廚師 The Saucier

2 湯水廚師 The Potager

3 魚料理廚師 The Poissoner

4 菜料理廚師 The Entremetier

5 燒燴料理廚師 The Rotisseur

6 煎炸料理廚師 The Grillardin

7 凍肉、頭盤料理廚師 The Garde Manger

8 甜品及包類廚師 The Patissier

9 替位廚師 The Tournant

10 助理廚師 Commis Cook

# I.2 廚房員工的權責
## Job Description

行政總廚：是廚房的靈魂人物，負責所有對外及對內之工作，監控整個廚房的運作、設計餐單、控制成本、與餐廳經理緊密聯繫等。

助理主廚：負責協助總廚日常之運作，監控所有生產過程、出品品質及員工培訓等。

部門主廚：負責各部門之主要日常運作，特別是生產管理。

1. 汁醬廚師：主要負責製作汁醬、湯底、熱頭盤及炒類料理。

2. 湯水廚師：主要負責湯底及湯類之料理製作。

3. 魚料理廚師：主要負責魚類及海鮮類之料理。

4. 菜料理廚師：負責蔬菜、澱粉質類如薯仔及蛋類之料理製作。

5. 燒燴料理廚師：負責燒製之肉類，先燒後燴之料理及肉汁之製作。

6. 煎炸料理廚師：負責所有煎或炸之肉類及海鮮料理製作。

7. 凍肉、頭盤料理廚師：主要負責所有凍肉及菜之料理，包括沙律、沙律汁、凍醬糕、凍頭盤及自助餐之頭盤料理製作。

8. 甜品及包類廚師：負責所有甜點及麵包類之製作。

9. 替位廚師：負責當其他廚師休息時代替其位置。

10. 助理廚師：初級廚師主要負責協助各部門之廚師製作料理及搬運貨品、清潔地方及用具等工作。

* 大多數的小型酒店及餐廳之中，因成本、人力及物力的關係，當然不能有如此精細的分工，故大多以一人身兼數職來運作。

**Executive Chef:** As the soul of a kitchen, the executive chef is responsible for managing internal and external affairs, monitoring the operation of the whole kitchen, designing menus, controlling the cost and keeping close contact with restaurant's managers.

**Sous Chef:** Responsible for helping the executive chef in daily operation, monitoring every production process and quality of products as well as training of staff.

**Chef de Partie:** Responsible for managing each department's daily operation, especially the production.

1) **The Saucier:** Responsible for making sauces, stocks, hot appetizers and sauté dishes.

2) **The Potager:** Responsible for making stock and producing dishes from broth.

3) **The Poissoner:** Responsible for producing dishes of fish and seafood.

4) **The Entremetier:** Responsible for producing dishes of vegetables, starch (like potatoes) and egg.

5) **The Rotisseur:** Responsible for making roast meat, dishes which are first roasted, and then braised as well as meat sauce.

6) **The Grillardin:** Responsible for pan-frying and deep-frying meat and seafood.

7) **The Garde Manger:** Responsible for all cold meat, vegetables dishes,including making salad, salad sauce, cold appetizer and appetizer for buffet.

8) **The Patissier:** Responsible for making desserts and bread.

9) **The Tournant:** When the chefs are on leave, the tournant will replace them.

10) **Commis Cooks:** A junior cook who is responsible for helping the chef of each department to prepare dishes, transport goods, clean the place and tools.

\* For many motels or restaurants, division of labor is not carried out in such a detailed way because of cost, manpower and resources. In many cases, one has to take different posts at the same time.

# II. 餐單設計
# Menu Designing

## II.1 設計餐單的考慮因素
## Factors Affecting Menu Design

餐單的主要目的是吸引及帶領人客去認識餐廳，從而選擇他們所想要的食品。要設計一份理想的餐單，當中所需之技巧頗不簡單，以下為一些須依從的原則：

1. **餐牌必須達到人客所要求**：令人客滿意乃整個進餐過程之首要目標，故此餐牌必需迎合人客的要求。

2. **餐單必須能跟隨市場之方向**：市場方向為大勢之所趨，即為大多數人客所想要之食品。

3. **餐單必須達至品質之要求**：餐單之設計必須根據餐廳本身之等級定位，由級數及價錢來訂定食品之品質，當然出品亦必須達到一定的水平。

4. **餐單之設計必須符合成本之要求**：每間餐廳均需為其食品作成本定位，成本額通常為銷售額之20-40%不等，視乎餐廳之定位等級，因為不同等級的餐廳所使用之材料亦大大不同。

5. **餐單必須真確**：即餐單中所列出之食品必須為真實用料，例如指明為新鮮的便需為新鮮之食品，不可使用雪藏品代替。

6. **出品項目必須為能力所及**：意指所有之出品必須考慮到廚房員工能否做到，樓面員工又能否達到奉客時之要求，例如你不能要求普通餐廳之員工出品法國頂級料理。

7. 工具設施之要求：設計餐單時，必須考慮到餐廳本身所擁有之設施及廚房有否足夠之工具去製作；例如一間咖啡室不會供應自助餐等。

8. 餐單之價錢：除需要考慮成本外，亦應根據餐廳之所在地及市場之定位來計算售價。

9. 餐單項目之平衡：在設計餐單時亦應考慮食物的平衡度，如上菜食物由輕至重，又如頭盤或湯由海鮮煮成，那麼大盤便可考慮為肉類，當然如能計算營養份量則更為理想。此外，在餐單設計中亦應避免使用重複字詞及用料。

A good menu aims to attract and lead guests to read then choose foods they want. It is not easy to design a good menu, here are some standards for your reference:

1) **Satisfy guests'needs:** The main aim of making a meal is to make guests satisfied.

2) **Follow market trends:** Market trends reflect what most guests want.

3) **With quality foods:** Menu should be designed according to the restaurant's class, the prices should reflect the quality standards of foods. Products should necessarily be of excellent quality.

4) **Meet the budget:** Every restaurant should set cost budget for its foods. Usually, the cost is 20-40% of the sales, depends on the grade of restaurant as ingredients used in different graded restaurants vary.

5) **Be true:** Foods mentioned in the menu should be accurate and true, for instance, fresh foods should not be replaced by frozen ingredients.

6) **Foods are able to be made:** Make sure that chefs in the restaurant are able to make dishes listed on the menu and staff are qualified to serve guests. For example, you cannot expect a general staff to cook a world-class French cuisine.

7) **Consider restaurant's facilities:** When you design a menu, you should consider the facilities and cooking tools in your kitchen, for example, you cannot provide buffet in a café.

8) **Price:** Other than cost, you should also consider restaurant's location and market trend to set the price.

9) **Make a balanced menu:** Make your menu a balanced one. Heavy dishes should be served after light foods; if appetizer or soup contains seafood, meat can be the main course. It will be ideal if nutrition values of foods are taken into account. Try to avoid repeating words or ingredients.

## II.2 餐牌之種類
### Types of Menu

餐牌之種類繁多，因應不同之形式而有所改變，如餐廳之形式（咖啡室、自助形式餐廳、主題餐廳、酒店上房服務等），又例如時間性的分別（早餐、午餐、晚餐等），還有出品的分別（甜品餐單、飲品餐單等）。（A, B）

另一個分類之概念為整體餐單之分類，例如（散餐 à la Carte menu，特定餐 set menu 等）。（C, D）

傳統中餐單之設計流程分為以下幾點：

1. 頭盤類：分凍及熱頭盤類，主要讓人客打開胃納，使他們更能享受隨之而來的美食。

2. 湯類：通常為第二道菜，亦有凍熱之分，但主要為熱類，其作用為增強進餐之氣氛及人客之味覺。

3. 沙律類：分類繁多，有蔬菜沙律、配肉沙律或海鮮沙律等；沙律可同時作為頭盤或大盤之用。

4. 大盤類：分有肉類、海鮮類及蔬菜類。大盤為全張餐單之主要食源，所以通常份量較大，很多時餐廳之主要招牌菜式均來自大盤類。

5. 甜品類：在整張餐單中甜品把守最後一關，在我來說是不可或缺的。如餐單中沒有甜品的話，就如故事沒有結尾一樣，總覺不是味兒。

A

B

There are various types of menus. They are set according to the types of restaurants (cafe, buffet restaurant, theme restaurant, room services, etc.), time (breakfast, lunch, dinner, etc.) and types of products (dessert, beverage, etc.) (A,B)

As a whole, menu can be divided into à la Carte menu or set menu. (C,D)

Traditional design of menu includes the following items:

1) **Appetizers:** There are hot and cold appetizers which can stimulate guests' appetites so that they can enjoy the coming dishes.

2) **Soups:** Usually soup is the second dish and can be served both hot and cold, but hot soup is more common. Soup can improve eating mood and guests' tastes.

3) **Salads:** There are many kinds of salads like vegetable salad, meat salad, seafood salad, etc. Salad can be served as an appetizer or a main course.

4) **Entrés (Main Course):** It can be classified as meat, seafood and vegetables. As the most important source of food in a menu, main course is generally large in portion and the signature cuisine of the restaurant.

5) **Desserts:** Dessert is the last part of the whole menu. To me, it is indispensable. If no desserts are served in the menu, it'll be a pity, just like a story without ending.

C

D

# III. 西餐的基本烹調方法
## Basic Cooking Methods of Western Cuisine

要做出美味可口的食物，廚師要懂得九種基本烹調法。

1. 炒：炒時要用強火，因為肉的蛋白質遇到火會凝固，所以用猛火可使肉的表層收縮，封住肉內的汁液不讓它流失。

2. 煎：是指食物以鑊及少量熱油來烹調，食物表面直接與熱油接觸。需要留意保留肉中汁液才能做出美味之菜式，可於調味後加上薄層麵粉來保護或以猛火煎封肉面均可。

3. 炸：是指食物全部接觸熱油來烹調。因為這種方法會使部份油脂在烹煮過程中滲入肉中，所以較適宜用於脂肪含量較低之材料；另外要留意的是油的溫度，油溫太低會引致過多油脂滲入食物之中，影響食味及效果；可沾上麵包糠或粉漿作為保護層，同時亦能將肉汁保留。

4. 烘烤：烤是指將食物放置於焗爐內烹調，食物是以空氣對流的方式傳導熱力。食物中一定之水份會流失，所以不適用於太小的肉類，所燒的肉類應最少為500克以上；開始時應用猛火，再轉為較低之火力去完成所需之熟度。

5. 燒烤/扒：此烹調法適用於較高質素及較薄之肉類，如西冷牛排；將肉食置於爐架上以爐火直接烹煮，如烹煮含較少脂肪的肉塊如牛柳時，便應掃上少量油脂以避免其乾焦，此方法能在較短之時間內煮熟食物，同時又能保存肉質本身的味道。海鮮亦能以此法烹調。

6. 燴：是指以一定之液體，用較長時間來烹煮食物，要訣是以強火煮開後再以慢火來燴。這種方法能有效地將食物的養份及香味保留。需要注意的是，應用較韌的肉來做，因肉味較濃而燴後肉件不會散開；同時應加上煲蓋烹調，以保留食物的水份及香味。

7. 蒸：以蒸氣來烹煮食物能充份地保存食物的養份。這方法製作出來的食物，清淡嫩滑，特別適合注重健康的人士，但缺點是需要較長之烹調時間。

8. 焓：以猛火及大滾的水或液體來烹煮食物，當水的溫度達至攝氏100℃，便會產生大量氣泡，此方法能在最短的時間內把食物煮熟；如蔬菜、意大利粉等。亦可用於果醬的製作中，這樣可將果醬內之水份除去。

9. 浸煮：是指將食物置於微滾之液體中烹煮，可使用水、湯或酒等；適用於肉質較嫩之食物，如海鮮或水果等。這方法能有效地保存食物的香味和嫩滑度。

A chef should manage nine basic cooking skills in order to make delicious dishes.

1) **Sauté:** Use high heat during sautéing. Protein in meat solidifies over high heat, when meat is cooked this way, the skin layer of meat contracts and hence meat juice is sealed and will not flow out easily.

2) **Pan Fry:** During pan frying, food is cooked with little hot oil in a wok or pan, surface of food touches hot oil directly. To make a luscious dish, coat meat with a thin layer of flour or fry meat over high heat so as to keep the juice.

3) **Deep Fry:** During deep frying, the whole piece of food is soaked in hot oil. In this way, some oil penetrates into the meat, so ingredients containing low fat is more suitable for deep frying. In addition, be aware of oil temperature. Taste of food will be affected if it is deep fried in low-temperature oil as much oil will penetrate into the piece of meat. In this case, coat meat with a thin layer of breadcrumb or flour to keep the juice.

4) **Roast:** It means putting the food in an oven, food is cooked by heat circulation. During roasting, water in food is lost. Meat lighter than 500g is not suitable for roasting. Start roasting with high heat, then switch to low heat for a better result.

5) **Grill:** It is good to grill high-quality or thin piece of meat like sirloin steak. Put meat on grid, cook over fire directly. Brush some oil on meat with less fat, like beef tenderloin, to prevent scorching. Grilling can cook meat within a short period of time and help to preserve its original taste. Seafood can also be grilled.

6) **Stew:** Stew is to cook food in liquid over a long period of time. Start to boil liquid with high heat then stew over low heat. By stewing, nutritions and fragrance of food can be kept. Use tough meat as its flavour is stronger and will not fall apart after stewing. To keep the moisture and aroma, cover with a lid during stewing.

7) **Steam:** Nutrition of food can almost be kept when steamed. Steamed food is light in flavour, tender and smooth and so it is good for people who prefer a healthy diet. Long cooking time is however a disadvantage.

8) **Boil:** Boil water or liquid over high heat, then put in food. The water temperature should be 100℃ , bubbles can be found when water is boiled. This method can cook food like vegetables or spaghetti in the shortest period of time. Jam can also be made this way as water in jam can be removed.

9) **Poach:** Poach means to cook food in slightly boiled liquid like water, soup, wine, etc. Poaching is suitable for food with soft texture like seafood or fruits, and can effectively preserve the fragrance and the tender texture of the meat.

第二課
廚房裏的秘密 Cooking Basics

# I. 工具概覽
## Cooking Tools Galleries

**❶ 量匙 Measuring Spoon:** 量度微少食材的量器，主要分為茶匙和湯匙。Small amount of dry ingredients are measured in teaspoons and tablespoons.

**❷ 量杯 Measuring Cup:** 量度液體的量器，每量杯約為250毫升（英式）。It is commonly used to measure liquid. In general, one measuring cup is 250ml (UK style).

**❸ 時間器 Timer:** 計算烹調時間的量度器。It is a time-counting machine to measure preparation time or cooking time.

**❹ 砧板 Cutting Board:** 用作處理食物的墊底工具，主要以木材或特殊膠質用料製造，以不損刀鋒和易於清理為佳，亦會按照不同材料或用途而選用不同的砧板。It is generally made of wood or poly-propylene and placed on the table to cut ingredients. Those do not blunt sharp knives and can be cleaned easily are your good choices. Choose different boards according to different ingredients and purposes to fit your usages.

**❺ 食物鉗 Tongs:** 以夾取食物或作烹調食物的器具，形如V字，屬不鏽鋼原料。They are used to pick up and transfer food. Its shape is just like "V" and it is made of stainless steel.

**❻ 湯杓 Ladle:** 以盛取湯汁，有不同尺碼，標準容量為200毫升。 They are used for holding liquid. There are different sizes, the standard capacity is 200ml.

**❼ 蔬菜刨 Vegetable Peeler:** 用作削去蔬果的外皮之用。It is used to peel vegetables and fruits.

**❽ 木匙/耐熱膠匙 Wooden Spoon / Heatproof Spoon:** 可作攪拌或作翻炒食材之用，避免刮花易潔鑊內的保護膜。It is used to stir or stir-fry ingredients. Will not scape away the protection of non-sticky frying pan.

**❾ 磨茸器 Grater:** 用作把食材磨碎成茸，分有不同尺碼和孔子以適應不同需要。It is used to mince ingredients. There are different sizes and perforations for different needs.

**❿ 廚房剪刀 Kitchen Scissors:** 用作修剪食物如海鮮和蔬菜等。They are used to trim foods like seafood or vegetables.

**⓫ 隔篩 Sieve:** 用作過濾食物、隔除雜質和篩粉之用。 They are used to strain liquid ingredients, separate wastage substances or sift flour.

**⓬ 切肉刀 Chef's Knife:** 以切割肉類之用，應揀選鋼質良好、刀鋒利的切肉刀。刀身長約15-25厘米為合。It is used to cut and trim meats. Good knives should be sharp, of 15-25-cm-long and made of good quality steel.

## I.1 廚刀分類
### Classification of Knives

廚刀是料理製作不可缺少的重要工具。廚師們會遵照安全守則使用廚刀，意即刀具必須常常保持鋒利，握刀手勢正確和切割食物時經常保持乾爽和清潔。以下介紹一個傳統廚房中必備之刀具：

Knives play a vital role in a kitchen and are applied in numerous tasks during cooking. In accordance to safety guidelines, the chefs should always keep their knives sharp, hold them in correct gesture for cutting, and store them in a clean and dry drawer to prevent the knives from dulling. For better understanding, there is a general introduction of common knives in a standard kitchen.

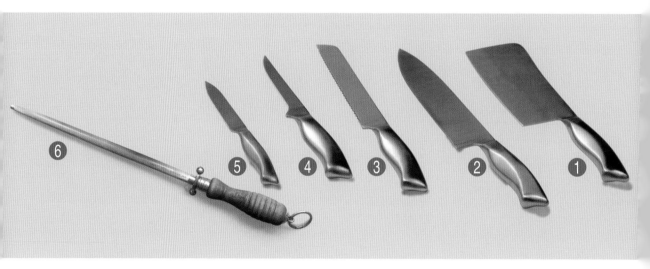

❶ 西式剁刀（菜刀）**Western Cleaver:** 刀身沉重，刀面呈長方形，闊平鋒利，適合砍骨和切割連骨肉塊。It is heavy enough for chopping bones and meat joints with a large and flat rectangular blade.

❷ 廚刀 **Chef's Knife:** 刀面呈三角形，長度約由 15 厘米至 30 厘米不等，弧形刀鋒，容易轉動刀身作切割之用。Its length ranges from 15cm to 30cm with long triangular-shaped blade. The chefs can rock the knife for easy cutting by using its curved edge.

❸ 鋸齒刀 **Serrated Knife:** 分有大、小兩款尺碼。小號刀子長約 13 厘米，適合切割蔬果；大號刀子可均勻地切出蛋糕片和麵包。There are two sizes. The small one is 13cm long which can cut fruit and vegetables neatly, while the larger knife can slice cakes and bread evenly.

❹ 去骨刀 **Boning Knife:** 全長約9-15厘米，刀鋒長而硬，前端細而尖，易於削肉。Has a long rigid blade, about 9-15cm long. It is curved to a fine sharp tip to make boning meat easier.

❺ 去皮刀/雕刻刀 **Small Paring Knife:** 它是最實用的一張刀，刀面只有6-9厘米闊。由於尺碼小巧，易於控制，適宜切蔬菜、水果和肉。it is the most useful knife in a kitchen. Length of its blade is only 6-9cm, therefore it is easy to control and best for cutting vegetables, fruits and meat.

❻ 磨刀棒 **Knife Sharpener:** 它是一條質感粗糙的長鐵棒，用以削利刀鋒。只要把長棒與刀子以斜傾45°上下推動便可。It is a long rod of coarse-textured steel which can sharpen a knife by grinding the blade edge against it at 45°.

## I.2 基本刀章法
## Basic Cutting Skills

以下介紹多種基本的切菜模式，常用於各種料理之中，現以甘筍作實例示範。

The following part is to introduce assorted cutting skills. Those skills are often applied in dish making. The demonstration will use carrot as an example.

**幼絲 Julienne:** 切割後，尺碼約為2毫米。After cut, the size is about 2mm.

**小粒 Brunoise:** 切割後的尺碼應為3毫米³，是傳統清湯常用配料的標準大小。The standard size is about 3mm³ after cut. It is suitable for cooking classic consommé.

**粒狀 Medium Dice:** 體積要求約為1厘米³。The standard volume is about 1cm³.

**粗粒 Large Dice:** 標準體積約2厘米³，可以切成不規則形狀。The standard volume is about 2cm³. Can be of irregular shapes.

**切塊 Chop:** 不規則大小之粗塊，多用於燴菜中。Irregular shapes are allowed and it is often applied in stewing / braising dishes.

# II. 材料的選購與烹調關係
## Good Ingredients, Better

在烹調領域中，即使學會了烹煮方法，如果材料不理想，仍然無法做出芳香美味的菜式。專業廚師通常會被認定是選擇和辨認材料的高手。不過，即使擁有豐富的知識和專業技術、以及具備足夠工具的廚師，要是沒有良好材料，也是徒然的。每當人們提及到良好材料，總會聯想到貴價物品，這是錯誤的想法。因為價格便宜、質地新鮮而味道良好的材料仍有很多的。

專業廚師和普通家庭主婦的烹調技巧其實有开很多相似的地方，在程序和調味上都相差不遠；當然，專業廚師多接受過嚴格的訓練，在知識與技巧上都有开一定的優勝之處，但最重要的一點是專業廚師能瞭解合乎材料的烹調法。舉例說：sirloin steak用來燒或煎時，效果相當好，因肉位於牛背上，有適量的運動及油脂，但如果用來煮湯或燴肉的話，那便不好吃了；相反，如果用腿肉來做湯或燴的菜式便相當理想，如腿肉用於燒烤或煎的話便太硬了。

所以就算是多麼便宜的材料，只要使用合適的烹調方法，便能做到美味的菜餚；而使用貴價材料是另一層次的烹調方法，這況提出的是基礎烹調的重點，為將來菜式的變化作根基。

Even if you have mastered cooking skills, you may still be unable to cook a delicious and fragrant dish if you don't have good ingredients. A professional chef is always honored as an expert of choosing and identifying ingredients. Ingredients are very important. Though you have excellent cooking knowledge, skills and tools, you still cannot make a luscious dish without good ingredients. Usually when people talk about good ingredients, they immediately think of the expensive ones. But this is a wrong concept. In fact, many inexpensive ingredients are fresh and delicious.

Professional chefs and ordinary housewives have similar cooking skills in regard to procedures and seasonings. But of course, professional chefs are strictly trained and they have better cooking knowledge and skills. Most importantly, they choose the right method to cook ingredients. For example, when cooking sirloin steak, which is the back of a cattle, it is best to be roasted and fried as it contains fat and this part is exercised very often. If it is cooked in soup or simmered, it will not be tasty. However, cattle's legs are good to make soup or be simmered.

As a result, with inexpensive ingredients and suitable cooking methods, you can still make a delicious dish. Expensive ingredients are used for advanced cooking. Based on the main points about cooking here, you can vary your dishes on your wish.

# III. 葡萄酒作為烹煮用途
# Cook with Grape Wine

在很多菜式的烹調中，都需要加入不同種類的酒，以使其香味溶於菜式之中。

我記得有一次在顧客前做了一道菜，有些女士卻對我說：「我若吃了含有酒精的食物，便會醉的。」我便告訴她們：「不用擔心！你們大可以放心的吃，這道菜中完全沒有酒精的！」她們嚐後都十分滿意，原因是我點火將菜餚中的酒精燒去，所以餘下的只是香味與精華。通常做料理時，酒精在烹煮時被蒸發了，吃後也不會醉倒。

西餐中使用最多的是白酒，它是做魚、海鮮、牛仔肉及雞菜式時不可缺的。除了能有效去除海鮮的腥味，還能帶出其原有的鮮味；做餐時使用較便宜的白酒便可，因用幾千元和幾百元的白酒做出同樣的菜式，得出的結果並不易辨別出。選購時應揀選乾性的白酒，太甜的白酒會影響菜式的原味。

馥郁的紅酒是牛肉的絕佳配搭，用作浸漬或烹煮均十分理想；紅酒多用於肉類料理，也適用於一些特殊之海鮮烹調中。

葡萄酒的用法一般分為以下幾種：

1. 將酒煮熱後加入湯汁之中，或直接把酒加入湯汁之中；
2. 用作浸漬食肉之用；
3. 將已煎或烤好的海鮮或肉類淋上葡萄酒；
4. 用於正在煮製的菜餚之中。

Many different kinds of wines are used during cooking so that fragrance of wine can be penetrated into foods.

I remembered what a lady told me when I cooked a dish for my guests, she said, "I'll get drunk if I eat food which is cooked with wine." Then I told her, "Don't worry, just eat whatever you want, this dish doesn't contain any alcohol." Once they had enjoyed the dish, they were all very satisfied as alcohol was removed during burning, only aroma of wine was left. During cooking, alcohol is usually evaporated, so dishes cooked with wine will not make you drunk.

White wine is mostly used in making western dishes. It is essential in cooking fish, seafood, veal and chicken. It cannot only remove unpleasant odor of seafood, but also release the freshness and original taste. It is good enough to use inexpensive white wine to cook as cuisine cooked with expensive wine and relatively cheap wine cannot be distinguished easily. Choose dry white wine, if the white wine is too sweet, original tastes of dishes will be destroyed.

Perfumed red wine is a perfect partner of beef, it is ideal for both soaking and boiling. Red wine is usually used in cooking red meat, but it can also be used to cook some special seafood.

Grape wines are generally used in the following cases.

1. Bring wine to the boil, add in sauce or add directly into sauce.
2. Soak meat in wine for use.
3. Sprinkle wine over fried or roasted seafood or meat.
4. Add in wine during cooking.

# IV. 食材概覽
## Grains, Pasta & Vegetables

### IV.1 穀物與麵概覽
Grains and Pasta Galleries

⑥ 野米 Wild Rice

⑩ 長身米（香米）Long-grain Rice(Jasmine Rice)

⑦ 印度米 Basmati Rice

⑨ 短身米（意大利米）Short-grain Rice(Risotto)

⑧ 粗麥 Cous Cous

④ 闊條麵 Tagliatelle

② 直通粉 Penne

⑤ 幼天使麵 Angel Hair Pasta

① 意大利粉 Spaghe

③ 螺絲粉 Fusilli

## IV.2 蔬菜概覽
## Vegetables Galleries

24 西蘭花 Broccoli

12 紅櫟葉生菜
Red Oak Leaf Lettuce

14 牛油生菜
Butter Lettuce

13 紅珊瑚生菜
Red Coral Lettuce

25 茄子 Eggplant

27 番茄 Tomato

16 苦白菜(捲葉苦苣)
Curly Endive

15 刁草 Dill

11 大蒜 Leek

18 百里香 Thyme

20 甜椒 Bell Pepper

23 辣椒 Chilli

17 羅勒 Basil

21 白菌(蘑菇)
Button Mushroom

26 意大利青瓜
Zucchini

芹菜 Celery

19 青瓜 Cucumber

22 鮮冬菇
Shiitake Mushroom

# V. 香草與香料的運用
## The Use of Herbs & Spices

### V.1 西餐的基本烹調方法
### Basic Cooking Methods of Western Cuisine

很多人認為香草與香料在烹調中很難調配，只要備有以下數種香料，便可以作料理的開始。

Many people think that herbs and spices are difficult to be matched up. If you have the following spices, you have already had a good start in seasoning ingredients.

❶ **胡椒 Pepper:** 分有白、黑、粉紅、青等不同顏色，各有不同之味道；但在大多數料理中，白和黑胡椒均已足夠，在料理中加入適量胡椒則有畫龍點睛之效。There are white, black, pink and green peppers. Different colors of peppers contain different tastes. Usually, white and black pepper powder is already enough for seasoning. Just some pepper can make dishes brilliant.

❷ **雜香料 Mixed Spices:** 能廣泛的應用於各種料理之中，無論湯汁、肉類、海鮮，甚至乎甜點均能應用得到。All spices can be widely used in different kinds of food such as sauce, meat, seafood as well as dessert.

❸ **香葉 Bay Leaf:** 能有效地去除異味，如雞湯之中的腥味等；切忌過多，因其香味十分強烈。It can remove odor like offensive smell of chicken soup. Don't add too much as it has a very strong smell.

❹ **百里香 Thyme:** 又名百搭香草，可用於各種料理之中，如湯汁、燴菜，並經常用於醃肉及海鮮料理中。It can be used in many foods like sauce and hodgepodge, always used in marinated meat and seafood.

❺ 迷迭香 **Rosemary:** 是羊肉料理的最佳配搭。因味道強烈，所以只需較少份量便行；可加入湯汁或燴菜中，亦可使用於醃肉、釀餡、燒豬或燒羊腿之中。It is a good partner of mutton. As it is strong in smell, just a small amount is to be used. It can be added in sauce, hodgepodge or marinated meat, filling, roasted pig or roasted mutton.

❻ 鼠尾草（洋蘇草）**Sage:** 常用於釀餡、牛仔肉或豬肉菜式之中，尤其適合於雞肉料理。It is usually used in filling, veal or pork, especially suitable for chicken.

❼ 意大利香草 **Oregano:** 顧名思義，常用於意大利料理之中，特別適用於燴燉之菜式。It is always used in Italian dishes, especially suitable for simmered dishes.

基本上擁有這些香草與香料已足夠派上用場了，因為在廚房的料理當中並不是一開始便甚麼都具備的，隨着製作料理的技術愈高，配料便自然地逐漸增多。緊記香草與香料的正確使用方式是僅用少許即可，因為其本義是要突出菜餚的香味而不是掩蓋主菜的味道，這點必須牢記。

Basically, having these herbs and spices in a kitchen are enough. A chef does not need to have every kind of sub-ingredients when he/she first cooks. When the chef becomes more skilled, he/she may need more sub-ingredients. Remember, don't use too much herbs and spices, a small amount is enough as they are used to give prominence to dishes' fragrance, they are not used to cover the tastes of main dishes.

# VI. 基本湯底
## Basic Stocks

### a. 香草束
### Bouquet Garni

香草束是任何西式湯底的靈魂，它包含了：大蒜、洋蔥、西芹、月桂葉和百里香等香料，縛成一束，放入湯煲內熬製而成，可提升湯底的真味。

Bouquet garni is the soul of all basic stocks, it includes leek, onion, celery, bay leaf and thyme. Bind them up and put in stock during cooking, this can bring out real taste of soup.

# b. 白湯底
## White Stock

容量：1公升
烹煮時間：2.5小時
Volume:1L
Cooking Time: 2.5 hrs

## 製法：

1. 所有材料放入煲內，慢火烹煮2.5小時，不時撤去湯上浮泡（A,B,C,D）。

2. 濾出清湯，用杓子壓出湯渣內的剩餘湯汁，待涼（E）。

3. 置入冰箱過一夜，移除湯上油脂（F）。

## Procedures:

1. Place all ingredients in pan and bring to the boil. Simmer for 2.5 hrs, skim foam frequently (A,B,C,D).

2. Strain the stock, press the solid with ladle to extract extra liquid and let cool (E).

3. Refrigerate stock overnight, then lift off any surface fat (F).

## 材料

雞骨或牛仔骨750克、洋蔥50克（粗切）、甘筍50克（粗切）、西芹50克（粗切）、丁香1粒、蒜頭2粒（切粒）、白胡椒粒10粒、香草束1束、清水1.5公升

## Ingredients:

750g Chicken or veal bone

50g Onion, roughly chopped

50g Carrot, roughly chopped

50g Celery, roughly chopped

1 pc Clove

2 pcs Garlic cloves, chopped

10 pcs White pepper corn

1 Bouquet garni

1.5 L Water

> ⊁Tips
> 放在冰箱內，可貯藏4日。
> Keep in refrigerator for 4 days.

# c. 黃湯底
## *Brown Stock*

容量：2公升
烹煮時間：4小時
Volume:2L
Cooking Time: 4 hrs

### 材料
牛骨1.5千克、番茄膏30毫升、洋葱100克(不去外皮，一開四)、甘筍150克(粗切)、大蒜80克(粗切)、西芹100克(粗切)、胡椒粒8粒、香草束1束、清水3公升

### Ingredients:
1.5kg Beef bone
30ml Tomato paste
100g Onion, unpeeled and quartered
150g Carrot, roughly chopped
80g Leek, roughly chopped
100g Celery, roughly chopped
8 pcs Peppercorn
1 Bouquet garni
3 L Water

### 製法：
1. 肉骨置焗爐以220℃火焗30分鐘，加入番茄糕拌勻後再焗15分鐘，待肉骨焗至淺黃，中途加入蔬菜拌勻(A,B)。
2. 放入少許水使焗盆的肉粒溶解，轉到湯鍋上，放入其他材料，慢火熬煮3.5小時(C)。
3. 濾出清湯，再從湯渣壓出水份，待涼(D)。
4. 置冰箱過一夜，撇去湯面油脂(E)。

### Procedures:
1. Roast meat bone at 220 ℃ for 30 minutes, add tomato paste, mix well, then roast for 15 minutes until lightly brown, add vegetables during roasting and stir to mix (A,B).
2. Deglaze with some water and then transfer to pan, add the remaining ingredients and simmer for 3.5 hours, skim often (C).
3. Strain the stock, press the solid with ladle to extract extra liquid and let cool (D).
4. Refrigerate stock overnight, then remove any surface fat (E).

### Tips
貯藏在冰箱內，可保存4日。
Tips: Keep in refrigerator for 4 days.

# d. 魚湯底
## *Fish Stock*

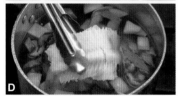

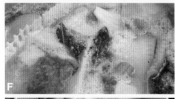

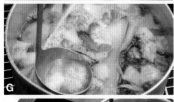

容量：2.5公升
烹煮時間：20分鐘
Volume:2.5L
Cooking Time: 20 mins

### 材料
魚骨2千克(修剪，粗切)、洋蔥100克(粗切)、甘筍100克(粗切)、西芹80克(粗切)、乾身白酒200毫升、白胡椒粒10粒、月桂葉2片、檸檬汁½個、清水3公升

### Ingredients:
2kg Fish bone, trimmed and roughly chopped
100g Onion, roughly chopped
100g Carrot, roughly chopped
80g Celery, roughly chopped
200ml Dry white wine
10 pcs White pepper corn
2 pcs Bay leaf
½ pc Lemon juice
3 L Water

### 製法：
1. 修剪魚骨，放進冷鹽水浸10分鐘。
2. 取出魚骨，修剪。連同其他材料一同放入煲內煮滾，再轉慢火熬煮20分鐘，撇去湯面泡沫(A,B,C,D,E,F,G)。
3. 用篩濾去湯渣，把杓子放在湯渣上，壓出多餘湯汁，待涼即成。

### Procedures:
1. Soak the bone and trim in cold salted water for 10 minutes.
2. Drain the bone and trim. Place all ingredients in pan, bring to the boil and simmer for 20 minutes, skim often (A,B,C,D,E,F,G).
3. Remove residue with sieve, put ladle on residue to extract excess soup. Let cool and serve.

### →Tips
1. 可貯藏在冰箱中4天。
2. 將魚骨浸於冷鹽水中，可去除血水和腥臭味。
3. 可選用白魚如鰈魚，或是粉紅魚如三文魚(即鮭魚)。
4. 熬煮清湯20分鐘以去除湯汁苦澀味。
1. Can be chilled for 4 days.
2. Soak the bone in cold salted water to remove blood and muddy taste.
3. Use white fish such as sole, or pink fish such as salmon.
4. Simmer stock for 20 minutes to get rid of bitter taste.

# e. 湯底的貯藏
## Storage of Stock

湯冰粒的製造：

1. 把原清湯倒在製冰格內，置於急凍櫃中4小時至凝固。

2. 凝固後的冰粒可作貯藏或應用。

3. 把冰粒從製冰格取出，放在膠袋內，回置冰箱備用(A,B,C)。

**Making stock cubes:**

1. Pour cold concentrated stock into ice-cube tray, then freeze for 4 hrs until solid.

2. When the cubes are frozen, they are ready for storage or use.

3. Remove them from tray and place them in plastic bag, return to freezer and use when needed (A,B,C).

# f. 油麵糊 / 麵撈
## Types of Roux

> **Tips**
>
> 麵撈的烹煮時間長短不一，視乎所需顏色而定。基本上，所有麵撈皆用上同量麵粉、牛油或菜油烹煮，所以濃稠度一般相同。至於用於做麵撈的汁液則有所不同：白汁用牛奶、高湯白麵撈用有味奶，而高湯則用來做傳統基本法國醬汁。
>
> Roux are cooked for different period of time, it depends on the colors needed. Usually, equal amounts of flour and butter or oil are cooked with all kinds of roux, so their thicknesses are generally the same. The sauces that are used to make the roux are different. Milk is used to make white sauce, infused milk for béhamel and stock for velouté.

**i) 白麵撈 ( 白色油麵糊 )：**

用以製作奶白汁和白汁，一般只煮1-2分鐘，亦可煮久一點令麵粉味道消除，麵糊仍未變色。

**i) White Roux:**

Used to make béhamel sauce and white sauce, usually it is cooked for 1-2 minutes or longer period of time to get rid of flour taste. Color still remains unchanged.

**ii) 高湯白麵撈：**

它是用以製作肉湯的基本醬汁，白色高湯以上雞、魚或小牛肉熬製而成。煮2-3分鐘直至麵撈呈淺金黃色。

**ii) Blond Roux:**

It is the traditional and basic velouté sauce, the white stock is made of chicken, fish or veal. Cook the roux for 2-3 minutes until it turns light golden in color.

**iii) 棕褐色麵撈：**

它是一種味道豐厚的麵撈，亦是傳統法式棕褐色醬汁的基本用料。這款麵撈以慢火加熱直至轉色，約需時5分鐘。

**iii) Brown Roux:**

This roux is rich in flavour and is the basic source of traditional French sauce-espagnole. It takes about 5 minutes to cook this roux over low heat until the color turns brown.

# VII. 醬汁與調味
## Precious Sauces and Dressings

## VII.1 不同醬汁的烹煮法
### Assorted Sauces

# *a.* 法式奶白汁
## *Béhamel Sauce*

容量：300毫升
Volume: 300ml

> →Tips
>
> 以上份量作食物淋面之用，如果要作覆蓋食物，則需要把牛油和麵份量加倍。
>
> The portion of sauce is served as pouring sauce. To cover the whole dish, double the amount of butter and flour.

材料

牛油15克、麵粉15克、牛奶300毫升

### Ingredients:

15g Butter
15g Flour
300ml Milk

製法：

1. 把麵粉加進牛油溶液，以慢火烹煮，期間不斷用木匙攪動成白麵撈，約需 1-2 分鐘(A, B, C)。

2. 離火，徐徐加入熱牛奶，攪入麵撈內混合(D, E)。

3. 回火煮滾，期間不斷攪拌，直至熬煮成濃稠適中的醬汁(F, G, H)。

### Procedures:

1. Add flour to melted butter, stir with wooden spoon over low heat to create white roux, it takes about 1-2 minutes (A, B, C).

2. Remove pan from heat, gradually add hot milk, beat constantly to blend with roux (D, E).

3. Bring to the boil, stir frequently, simmer to the right thickness (F, G, H).

# 6. 高湯白汁

## Velouté Sauce

材料

牛油25克、麵粉25克、高湯500毫升

**Ingredients:**

*25g Butter*
*25g Flour*
*500ml Basic stock*

製法：

1. 把麵粉加進牛油溶液，以慢火烹煮，期間不斷用木匙攪動成白麵撈，約需2-3分鐘(A, B, C, D)。

2. 離火，讓其稍稍降溫。

3. 不斷攪拌，徐徐加入熱雞湯、牛肉湯或魚湯(E)。

4. 回火煮滾，不斷攪拌及以慢火熬煮10-15分鐘，期間不斷撇去面層浮泡(F, G)。

**Procedures:**

1. Add flour to melted butter, stir with wooden spoon over low heat for about 2-3 minutes to create white roux (A, B, C, D).

2. Remove from heat and let cool slightly.

3. Stir constantly; gradually add in hot chicken stock, veal stock or fish stock (E).

4. Bring to the boil, stir continuously and simmer for 10-15 minutes, skim the foam frequently during cooking (F, G).

# c. 基本黃汁

## Basic Brown Sauce

容量：750毫升
Volume: 750ml

### 材料
牛肉或小牛肉湯底1公升、葛粉35毫升、清水120毫升

### Ingredients:
1 L Brown beef or veal stock
35ml Arrowroot
120ml Water

### 製法：

1. 牛肉湯煮20分鐘，直至湯底份量和味道濃縮。
2. 葛粉和清水弄成糊狀，倒入已煮沸的清湯內，不斷用力攪拌。
3. 待汁煮至濃稠，離火，撇去面層浮泡。

### Procedures:

1. Boil brown stock for 20 minutes until stock volume and flavour condensed.
2. Mix the arrowroot and water to form a paste, pour into boiling stock, whisk constantly.
3. Cook the sauce until thickened, remove from heat and skim the foam.

註：這款汁醬常與燒烤肉或野味伴吃。它可以增加牛肉湯的色澤和味道，絕對是肉湯的精華。

**Remarks:** This sauce is commonly served with roasted meat or game. It can enrich the beef or veal stock's color and flavour, in fact, this is the essence of meat soup.

## VII.2 不同的調味醬
## Assorted Dressing

# *a.* 牛油醬汁 - 荷蘭汁

容量：250毫升
Volume: 250ml

## *Butter Sauce-Basic Hollandaise*

**製法：**

1. 打拂蛋黃和熱水，並以十分慢火和不斷攪拌約2分鐘至濃稠，離火(A)。
2. 徐徐加入暖和牛油清，順時針方向攪拌(B, C, D, E)。
3. 一邊攪打，一邊加入檸檬汁，再加入鹽和鮮磨胡椒調味(F, G, H)。

**Procedures:**

1. Whisk egg yolks and hot water over very low heat for about 2 minutes to ribbon stage. Turn off the heat (A).
2. Add warm clarified butter gradually until done. Whisk clockwise and constantly (B, C, D, E).
3. Slowly whisk in the lemon juice, add in salt and fresh pepper to season (F, G, H).

**註：** 這是一種以牛油和蛋黃乳化作用造成的醬汁，必須即做後而醬汁還保持暖和時享用。

**Remarks:** This is a light and butter based sauce made by an emulsion of eggyolk and butter. Serve the freshly sauce-made while it is still warm.

**材料**
蛋黃3隻、熱水45毫升、無鹽牛油清175克、檸檬½個(榨汁，隔渣)、鹽和鮮磨胡椒少許

**Ingredients:**
*3 pcs Egg yolks*
*45ml Hot water*
*175g Unsalted butter (clarified)*
*½ pc Lemon juice without residue*
*+ Salt, fresh pepper from the mill*

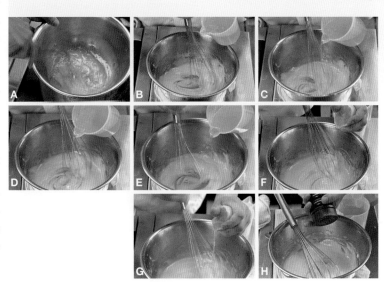

> ⇥Tips
> 如何製作牛油清？把牛油用慢火溶化，不要攪拌，離火，撇去表面雜質。小心用鋼匙取出牛油清，讓牛油的奶固體和雜質留在煲底。
> How to clarify butter? Melt butter slowly over very low heat without stirring, remove from heat and skim the surface. Spoon the butter into a bowl carefully to leave the milky sediments in the pan.

# 6. 蛋黃醬汁

## *Mayonnaise*

容量：320毫升
Volume: 320ml

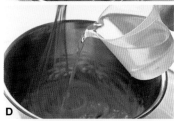

### 材料

蛋黃1隻、法國芥辣醬15毫升、橄欖油150毫升、菜油150毫升、白酒醋10毫升、鹽、鮮磨胡椒少許

### Ingredients:

1 pc Egg yolk
15ml Dijon mustard
150ml Olive oil
150ml Vegetable oil
10ml White wine vinegar
+ Salt, fresh pepper from the mill

### 製法：

1. 將一玻璃碗置於布巾上，穩定位置。
2. 把蛋黃、芥辣和調味料打拂至材料混合(A)。
3. 每次逐少加入橄欖油直至混合物濃稠(B, C)。
4. 慢慢加入菜油(D)。
5. 徐徐加入少量白酒醋，逐次攪勻直至完全混合(E)。
6. 調味享用。

### Procedures:

1. Place a deep bowl on a towel to prevent a slippery ground.
2. Whisk egg yolk, mustard and seasonings in the bowl until combined (A).
3. Whisk in olive oil gently until mixture thickens (B, C).
4. Add vegetable oil in a thin stream (D).
5. Slowly add in vinegar, whisk well after each addition to ensure it is completely mixed (E).
6. Season to taste.

**註：**軟滑的蛋黃醬是一種簡易的乳化作用醬汁，無論用手或用機攪拌醬汁，效果都是一樣的。

**Remarks:** A soft and creamy mayonnaise is a simple emulsion of egg yolk, vinegar, seasonings and oil. You can combine the ingredients either by hand or by machine, the result is the same.

### →Tips

1. 把所有材料和用具置於室溫下。切勿在倉猝下調醬。
2. 材料太冷、用具有油、混合材料太快都會令醬汁分離或不能混合，這時抽出少量蛋黃醬和15毫升酒醋混和，再把其餘醬汁逐少加入至完成。

1. Place all ingredients and equipments at room temperature. Do not rush when making mayonnaise.
2. The sauce will fall apart or curdle if the ingredients are too cold, the equipment has an oily surface or if the speed of mixing is too fast. When you find the sauce under curdiness, you can mix 15ml wine vinegar with a litte amount of mayonnaise until completely mixed.

# c. 油醋汁
## *Vinaigrette*

容量：130毫升
Volume: 130ml

材料
醋30毫升、法國芥辣醬10毫升、橄欖油90毫升、鹽、鮮磨胡椒少許

Ingredients:
30ml Vinegar
10ml Dijon mustard
90ml Olive Oil
+ Salt, fresh pepper from the mill

製法：

1. 把醋、芥辣和調味料混合，拂打至濃稠(A, B)。
2. 一邊徐徐加入油，不斷拂打，即成滑溜溜而濃稠的調味醬汁(C, D)。

Procedures:

1. Mix vinegar, mustard and seasonings together, whisk to combine and until thicken (A, B).
2. Slowly add in oil, whisk constantly until the dressing is smooth, thickened and well blended (C, D).

# VIII. 麵糰與麵條的製作
# Making Pasta Dough, Rolling and Cutting Pasta

## *a.* 麵糰製作

### *Making Pasta Dough*

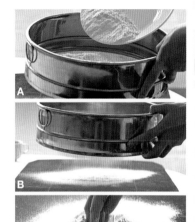

### 材料
高筋粉（麵包粉／強力粉）300克、雞蛋3隻（輕稍拂）、橄欖油15毫升、鹽少許

### Ingredients:
*300g Strong plain white flour*
*3 pcs Eggs, slightly beaten*
*15ml Olive oil*
*+ Salt*

### 製法：
1. 麵粉篩在檯上，開洞（A, B, C）。
2. 把雞蛋、鹽和橄欖油加於麵粉洞中（D, E）。
3. 用手指混合材料，直至雞蛋完全被麵粉吸收（F, G）。
4. 搓揉成軟滑而具彈性的麵糰，需時約10分鐘（H）。
5. 蓋布等待1小時，讓麵糰起筋。

### Procedures:
1. Sift the flour on a clean surface and make a large hole in the center (A, B, C).
2. Add egg, salt and olive oil to the hole (D, E).
3. Mix ingredients in the well with finger until eggs are completely absorbed by the flour (F, G).
4. Knead the dough until smooth and elastic for about 10 minutes (H).
5. Rest the dough and cover for 1 hour.

### ⇢Tips
手做麵條製作技巧簡單，只需花一點時間便會有好的結果。其技巧是利用陰柔力度搓麵糰，加上手溫碰觸，便能搓出軟滑具彈性的麵糰。選用上等的硬麥麵粉或粗麵粉是最佳做麵材料，不過該等材料不容易找到，以及處理困難，因此退而求其次，這個食譜便選用麵包粉了。

Making pasta by hand is simple, it only takes a little time and yet the product is good. Gentle kneading and the warmth of your hand both helps to create an elastic dough. The best flours are durum wheat and semolina flour, but they are not easy to proaire and it is also very hard to handle, so strong plain flour is used in this recipe as a substitute.

# 6. 麵條製作
## *Rolling and Cutting Pasta*

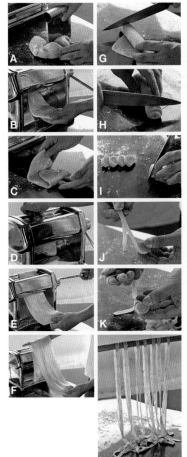

材料
*已鬆筋的麵糰400克*

### Ingredients:
*400g Rested pasta dough*

製法：

1. 把麵糰一分為四，把其中一份放在碾麵機內，調校碾碌至最寬位置(A, B)。

2. 每片麵皮碾開後，再摺成3褶，再次碾開，連續不摺疊麵皮，調校碾麵器，由厚至薄滾碾3次麵皮(C, D, E, F)。

3. 待麵皮薄如紙時，便可掛上灑上麵粉的木棍上，待10-15分鐘至乾身。

### Procedures:

1. Cut the dough into 4 pieces, feed one piece of dough in pasta machine, set the rollers at the widest point (A, B).

2. Fold each piece of dough into three folds and roll again, repeat three times without folding, reduce the width each time. (C, D, E, F)

3. When it is thin, hang the sheet of dough over a floured wooden handle. Let dry for 10-15 minutes.

### i) 手切麵條

1. 把微乾麵糰捲成鬆鬆的長圓柱體(G, H)。

2. 把圓柱形麵條均勻地切成小塊(I, J, K)。

3. 將切開的麵條掛起，風乾½小時。

### i) Cutting by hand:

1. Roll the dried dough into a loose but even cylinder (G,H).

2. Cut the cylinder crosswise into strips (I,J,K).

3. Hang the cut pasta over a floured wooden handle, let dry for ½ hr.

### ii) 機切麵條

1. 把麵皮放在麵粉布上，直至所有麵糰已捲起。

2. 放麵糰於機內，調校切割器，切出屬意的麵形。

3. 放置已切割的麵條於碾棍上，晾乾½小時。

### ii) Cutting by machine:

1. Place the sheets on a floured towel until all the dough has been rolled.

2. Place the dough in the machine, adjust and cut into the shapes you want.

3. Hang the pasta over a floured wooden handle, let dry for ½ hr.

> ⇢Tips
>
> 麵可以用手或用機碾成薄麵皮，晾至略乾，便可做成不同麵形。
>
> Pasta can be rolled into sheet by hand but it is quicker to use a machine and the pasta cut will be thinner. After a short drying time, the pasta sheet can be cut by hand or the pasta machine.

# IX. 馬鈴薯蓉與中東粗麥粉
## Mashed Potatoes & Cous

## *a.* 馬鈴薯蓉
### *Mashed Potatoes*

4人
4 pax

### 材料
馬鈴薯4個、牛油30毫升、忌廉40毫升、鹽、鮮磨胡椒少許

### Ingredients:
4 pcs Potato

30ml Butter

40ml Cream

+ Salt, pepper from the mill

### 製法：
1. 馬鈴薯去皮，用水煮至軟身，瀝乾(A)。
2. 趁熱把馬鈴薯磨成泥(B)。
3. 加入牛油，徐徐加入忌廉，攪打至輕薄鬆軟，加入適量調味(C, D, E)。

### Procedures:
1. Peel and cook potatoes in water until soft, drain (A).
2. Mash potatoes when it is still hot (B).
3. Add butter and gradually add in cream. Beat until light and fluffy, season to taste (C, D, E).

→Tips

柔軟煮熟的蔬菜是肉、禽、魚最好的配菜，為沙律帶來對比的顏色和口感。馬鈴薯可壓成軟綿和粗厚的薯蓉。要做好最佳薯蓉，就要選擇較佳的粉狀馬鈴薯。

Softened, cooked vegetables are popular partners of meat, poultry and fish as they offer contrast in colour and texture. They can be pressed into a smooth or coarse purée. To get the best mashed potatoes, you need to select the right type of floury potato, such as: - King Edward, it has white skin with pink patches, floury in texture.

# 6. 粗麥粉
## *Cous cous*

材料
中東粗麥粉125克、熱水250毫升、
牛油(室溫)30克

Ingredients:
125g Cous cous
250ml Hot water
30g Butter (room temp.)

製法：

1. 把蒸粗麥粉放入已抹了一層薄牛油的鍋況(A, B)。
2. 加入熱水，攪勻(C)。
3. 用中火煮5-8分鐘，離火，倒入牛油，攪拌至穀殼分離(D, E)。

Procedures:

1. Put cous cous in a lightly buttered pan(A, B).
2. Add hot water and stir until well-blended(C).
3. Cook cous cous in medium heat for 5-8 minutes, remove from heat and stir in butter, stir and separate the grain (D, E).

>Tips

1. 在北非，cous cous grain 原是一道菜名，這道菜是在一種叫作 Couscousiere 的特殊鍋子裏做的辣味肉和蔬菜。
2. 大多數粗麥粉都是先煮熟的，只需再加水蒸過便可。
3. 可以放入切碎的堅果、果脯或新鮮香草以豐富此菜式。

1. In North Africa, the cous cous grain is used to be the name of a spicy dish of meat and vegetable cooked in a special pot called Couscousiere.
2. Most cous cous is pre-cooked, only moistening and steaming are needed.
3. This recipe can be richer, you can serve it as a side dish, mix with chopped nuts, dried fruit or fresh herb.

# 第三課

## 繁忙的廚房料理製作

Cooking Workshops

| | |
|---|---|
| 蛋白3隻 | 3 pcs Egg white |
| 檸檬汁20毫升 | 20ml Lemon juice |
| 雜菜粒350克<br>（甘筍、西芹、洋葱、大蒜） | 350g Mixed vegetable cubes<br>(carrot, celery, onion, leek) |
| 免治牛肉350克 | 350g Minced lean beef |
| 暖牛肉湯底3公升 | 3 L Warm beef stock |
| 鹽少許 | + Salt |

製法 *Procedures*

**1** 蛋白打至呈泡沫。

**3** 將蛋白混合物加入暖湯，煮至滾沸。

**2** 把檸檬汁、雜菜粒和牛肉碎加入蛋白中，攪至完全混合。

**4** 一邊攪動肉湯，一邊熬煮約5分鐘，湯面呈現許多浮物雜質。

**5** 在浮面雜物中開一洞，用慢火烹煮1小時。

**1** Whisk egg whites until frothy.

**2** Add lemon juice, vegetable cubes and beef to the egg white and then mix well.

**3** Add egg mixture to warm stock and bring to boil.

**4** Whisk the beef stock for about 5 minutes until a crust forms.

**5** Make a hole in the crust and simmer for 1 hour.

**6** Break the crust to sieve the soup, then serve.

**7** Season with white.

**6** 弄破浮面雜質，濾出清湯。

**7** 以鹽調味。

# *Beef Consomm*

# 牛肉清湯

→Tips

1. 熬湯時，不要攪拌。

2. 加入些厘酒，以增加食味。

1. Do not stir when simmering the soup.

2. Add some dry sherry to make the soup more tasty.

| | |
|---|---|
| 牛油70克 | 70g Butter |
| 洋蔥90克（粗切） | 90g Onion, roughly chopped |
| 牛肉黃湯底500毫升 | 500ml Beef brown stock |
| 乾身白酒200毫升 | 200ml Dry white wine |
| 香草束1束 | 1 Bouquet garni |
| 古魯耶芝士（瑞士芝士）100克（刨碎） | 100g Gruyere cheese, grated |
| 鹽、鮮磨胡椒少許 | + Salt, pepper from the mill |

## 製法 *Procedures*

1. Sauté the onion in butter for about 20 minutes until caramelized .
2. Add in stock, wine and bouquet garni, bring to the boil, cover and simmer for 30 minutes.
3. Remove bouquet garni and season to taste.
4. Sprinkle the cheese on soup and put under hot grill for 2 minutes until golden.

**1** 把洋蔥和牛油炒20分鐘至深啡色。

**2** 注入清湯、白酒和香草束煮滾，加上煲蓋慢火熬煮30分鐘。

**3** 取出香草束，調味。

**4** 撒上芝士碎，置燒爐烘面2分鐘呈金黃色。

*French Onion Soup*
# 法國洋葱湯

→Tips

洋葱需以中細火慢慢炒至軟身及金黃色，讓它的香甜味能充份發揮出來。

Sauté the onion at medium low heat until soft and golden brown. It helps to release the flavour and sweet taste.

## 材料 *Ingredients*

| | |
|---|---|
| 甘筍450克（去皮，切丁） | 450g Carrot, peeled and diced |
| 洋蔥50克（去皮，切丁） | 50g Onion, peeled and diced |
| 薯仔（馬鈴薯／土豆）100克（去皮，切丁） | 100g Potato, peeled and diced |
| 牛油50克 | 50g Butter |
| 雞湯1公升 | 1 L Chicken stock |
| 香草束1束 | 1 Bouquet garni |
| 忌廉50毫升 | 50ml Cream |
| 鹽、鮮磨胡椒少許 | + Salt, pepper from the mill |

## 製法 *Procedures*

**1** 雜菜以牛油炒4分鐘至軟身，期間不斷翻炒。

**2** 注入清湯，放入香草束，熬煮20分鐘至蔬菜十分軟身。

**1** Sauté the vegetables with butter for 4 minutes until soft, stir frequently.

**2** Add stock and bouquet garni, simmer for 20 minutes until the vegetable is very soft.

**3** Remove the bouquet garni. Mix the soup in blender, sieve and then reheat in a clean pan with cream, season to taste.

**3** 取出香草束，倒入攪拌機內攪碎，濾出湯汁，回火，加入忌廉，調味即成。

*Puréed Carrot Soup*

# 甘筍茸湯

→Tips

1. 雜菜煮至十分軟身，使之容易攪碎。

2. 把乳酪（優格）放於湯面，豐富湯味。

1. Cook the vegetables until very soft for the ease of blending.

2. Add some plain yoghurt on top to enrich the taste of soup.

## 材料 Ingredients

| | |
|---|---|
| 連皮雜魚柳1千克(切大塊) | 1kg Mixed fish fillet with skin, cut into large pieces |
| 蠔6隻(留汁) | 6 pcs Oyster, keep the juice |
| 橄欖油50毫升 | 50ml Olive oil |
| 洋蔥100克(切碎) | 100g Onion, finely chopped |
| 西芹條80克(切碎) | 80g Celery stick, finely chopped |
| 紅甜椒100克(切粒) | 100g Red bell pepper, diced |
| 青甜椒50克(切粒) | 50g Green bell pepper, diced |
| 蒜頭2粒(切碎) | 2 cloves Garlic, finely chopped |
| 乾辣椒碎少許 | + Some dried chilli flakes |
| 藏紅花少許 | + Pinch of saffron strands |
| 番荽30毫升(切碎) | 30ml Parsley, chopped |
| 月桂葉1片 | 1 pc Bay leaf |
| 百里香葉少許 | + Some Thyme leaves |
| 番茄500克(去皮，去籽，切碎) | 500g Tomato, skinned, deseeded and chopped |
| 魚湯1公升 | 1 L Fish stock |
| 連尾中蝦500克 | 500g Prawns with tail |
| 黑朗姆酒20毫升 | 20ml Dark Rum |
| 鹽、鮮磨胡椒少許 | + Salt, pepper from the mill |
| 百里香葉8毫升(切碎)(用作裝飾) | 8ml Thyme leave, chopped for decoration, 12 pcs Croutes |
| 脆多士12片 | |
| 蒜頭甜椒醬60毫升 | 60ml Rouelle |

## 製法 Procedures

**1** 用橄欖油炒香洋蔥、西芹、甜椒、蒜頭和辣椒碎，加入藏紅花慢煮5分鐘，間中翻炒數下，直至雜菜軟身和呈淺褐色。

**3** 湯汁煮滾，加入實肉魚塊以慢火煮5分鐘，期間攪動2次；加入軟身魚塊煮5分鐘，再加入蝦後多煮2分鐘，最後加入蠔與蠔水煮1分鐘。

**4** 熄火，取出月桂葉，加黑朗姆酒和調味，放百里香作裝飾。

**5** 可獨立配上蒜頭甜椒醬或直接放湯中，又或放在脆多士上。

**2** 加入香草和番茄，注入魚湯，調味，煮滾，然後以慢火熬煮20分鐘。

**1** Sauté the onion, celery, bell pepper, garlic, chilli flakes with olive oil until flavour comes out, add saffron and gently cook for 5 minutes, stir occasionally until all vegetables are soft and lightly brown.

**2** Add herbs, tomatoes and fish stock, season, bring to the boil and simmer for 20 minutes.

**3** Bring the liquid to boil and place the firm fish in the soup first and simmer for 5 minutes, stir gently twice. Add soft fish and cook for 5 minutes, add in prawns, cook for 2 more minutes, add oysters and their juice and cook for 1 minute.

**4** Turn off heat and discard bay leaf, stir in dark Rum and season to taste. Sprinkle thyme leaves on the soup .

**5** Can serve Rouille separately or add to the soup or spread onto croutes.

# *Creole Bouillabaisse with Rouille Sauce*
# 海鮮大燴配蒜頭甜椒醬

>Tips

1. 盡量選取優質魚肉。

2. 法國脆多士製法：用烘烤爐烘焙中等厚度的法包片至金黃色，翻轉再烘焙另一面。

1. Use the best and freshest fish you can buy.

2. How to make croutes: toast medium-thick baguette under a hot grill until golden brown, turn over and toast the other side.

註：如果加入辣椒、胡椒和朗姆酒，便是具印度口味的海鮮湯。

Remarks: This classic Proven al seafood soup will have a touch of Indian taste if some chilli, pepper and rum is added.

| | |
|---|---|
| 牛油50克 | 50g Butter |
| 煙肉60克（切小粒） | 60g Bacon, diced |
| 洋葱100克（去皮切丁） | 100g Onion, peeled and diced |
| 魚上湯900毫升 | 900ml Fish stock |
| 馬鈴薯260克（去皮切丁） | 260g Potato, peeled and diced |
| 忌廉180毫升 | 180ml Cream |
| 麵粉15克 | 15g Flour |
| 海鮮雜錦（蝦，魚，扇貝）300克（切粒） | 300g Mixed seafood (shrimp, fish, scallop), diced |
| 番荽20毫升（切碎） | 20ml Parsley, chopped |
| 鹽、鮮磨胡椒少許 | + Salt, pepper from the mill |

## 製法 *Procedures*

**1** 用牛油將煙肉和洋葱拌炒3分鐘。

**3** 將忌廉和麵粉混和，倒入湯中。拌勻，在中細火上煮8分鐘，不時地攪拌。

**2** 加入上湯和馬鈴薯，煮5分鐘直至馬鈴薯軟身而不碎裂。

**4** 加入海鮮和適量的調味料，煮透，但不要煮過熟。

**5** 撒上番荽，與多士和薄脆餅乾配合食用。

**1** Sauté bacon and onion in butter for 3 minutes.

**2** Add stock and potatoes, cook for 5 minutes until the potatoes are soft but unbroken.

**3** Mix cream with flour and pour into soup. Mix well, cook over medium heat for 8 minutes, stir frequently.

**4** Add seafood, season to taste and heat thoroughly but do not over-cook.

**5** Sprinkle on top with parsley and serve with toast or crackers.

# *England Seafood Chowder*
# 英式海鮮周打湯

## ✦Tips
1. 不要把洋葱和煙肉燒焦。用中火煮至軟身和有香味發出便可。
2. 選用新鮮海鮮入饌，不要煮得過熟。
1. Don't brown the onion and bacon, cook it under medium heat until onion is soft and flavour comes out.
2. Use fresh seafood and do not overcook.

## 材料
## *Ingredients*

| | |
|---|---|
| 羅馬菜(長葉萵苣)600克 | 600g Romaine lettuce, cleaned and drained |
| (清洗及去除水份) | 180ml Caesar dressing |
| 凱撒沙律醬180毫升 | ½ pc French baguette, diced |
| 長法包½條(切丁) | 1 clove Garlic, crushed |
| 蒜頭1瓣(切碎)、橄欖油60毫升 | 60ml Olive oil |
| 帕爾梅森硬奶酪碎160克 | 160g Parmesan crisp |
| 鹽、鮮磨胡椒碎少許 | + Salt, pepper from the mill |

### 凱撒沙律醬料
### Caesar Dressing

| | |
|---|---|
| 蛋黃2個 | 2 pcs Egg yolk |
| 法式芥末醬10毫升 | 10ml Dijon mustard |
| 香醋30毫升 | 30ml Balsamic vinegar |
| 橄欖油250毫升 | 250ml Olive oil |
| 巴馬芝士40毫升(磨碎) | 40ml Parmesan cheese, grated |
| 鯷魚3塊(切細) | 3 pcs Anchovie, chopped finely |
| 蒜瓣3個(切細) | 3 cloves Garlic, chopped finely |
| 鮮磨胡椒少許 | + Pepper from the mill |

## 製法 *Procedures*

### 凱撒沙律醬

**1** 燒熱油,將麵包和蒜碎一起煎至金黃色,取出蒜碎。

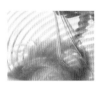

**1** 把蛋黃、芥末和醋攪在一起。

**2** 慢慢加入橄欖油,不停地攪拌,直至調味料均勻柔滑。

**3** 加入芝士、鯷魚和大蒜,拌勻,調味。

**4** 巴馬脆芝士:把磨碎的巴馬芝士均勻地撒在燒熱的平底鍋上,讓其軟化並融合在一起,熄火,放涼,芝士變脆,與沙律配合食用。

**2** 生菜切成大塊,用鹽和胡椒調味。

**1** Heat oil and cook the bread with garlic until golden, discard garlic.

**2** Slice the lettuce in big pieces, season with salt and pepper.

**3** Toss the lettuce with dressing, sprinkle croutons on top and serve with parmesan crisp.

**Caesar Dressing**

**1** Whisk the egg yolk, mustard and vinegar together.

**2** Slowly add in oil, whisk constantly until dressing is smooth.

**3** 沙律醬與生菜拌勻,撒上油炸的小麵包塊,與巴馬芝士一起食用。

**3** Add cheese, anchovies and garlic, mix well and season to taste.

**4** Parmesan crispy: Sprinkle the grated Parmesan cheese on hot pan evenly, let it melt and hold together, remove from heat and let it cool and be crispy, serve with the salad.

# *Caesar Salad*
# 凱撒沙律

➢Tips

1. 這道傳統菜式是用生雞蛋黃來做，如果你不喜歡用生蛋黃，可以用2湯勺蛋黃醬代替。

2. 油炸小麵包塊最好在做好後的半小時內撒到沙律上。

1. This traditional recipe uses raw egg yolk, if you do not prefer raw yolk, replace it by 2 tablespoons of mayonnaise.

2. The croutons are best made no more than half an hour before assembling with the salad.

## 材料 *Ingredients*

**6 pax**

| | |
|---|---|
| 雜錦沙律菜180克 | 180g Mixed salad greens |
| 橙3個(去皮，分開瓣) | 3 pcs Orange, peeled and segmented |
| 青瓜250克(切薄片) | 250g Cucumber, thinly sliced |
| 牛油果1½個(去皮，切片) | 1½ pcs Avocadoe, peeled and sliced |

### 沙律醬料 / *Salad dressing*

| | |
|---|---|
| 橄欖油125毫升 | 125ml Olive oil |
| 橙汁60毫升 | 60ml Orange juice |
| 糖30毫升 | 30ml Sugar |
| 檸檬汁30毫升 | 30ml Lemon juice |
| 香葱2棵(切碎) | 2 pcs Chive, chopped |
| 橙皮(磨碎)、鹽 | + Grated orange peel, salt |

## 製法 *Procedures*

**1** 沙律醬：把所有調味配料混和，拌勻。

2A

2B

**2** 牛油果去皮去核切片。

3A

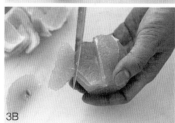

3B

**3** 橙去皮分開瓣。

4A

4B

**4** 把橙、青瓜和牛油果排放在盤子上，澆上沙律醬。

**1** To make the dressing: Combine all ingredients together, whisk well to blend.

**2** Peel, seeded and slice avocado.

**3** Peel and fillet orange.

**4** Arrange oranges, cucumbers and avocados on plate, spoon dressing over salad.

# *Avocado and Orange Salad*

# 牛油果香橙沙律

→Tips

一道漂亮美味的沙律，必須採用新鮮材料，保持潔淨和乾爽，而加入水果可增加新鮮味道。

A beautiful and delicious salad should use fresh ingredients, clean and dry. Add fruits in the salad to enrich the flavour of freshness.

新馬鈴薯500克
雞蛋3隻
青豆角180克（修完整）
綠甜椒100克（去籽，切片）
黑橄欖120克
番茄300克（切成角形）
青瓜180克（切厚角）
青葱3棵（切成1寸長段）
新鮮吞拿魚柳800克
鹽、鮮磨胡椒碎少許

吞拿魚沙律醬料
檸檬汁100毫升
蒜頭1瓣
橄欖油300毫升
香葱1克（切碎）
鹽、鮮磨胡椒少許

500g New potato
3 pcs Egg
180g Green bean, trimmed
100g Green bell pepper, seeded and sliced
120g Black olive
300g Tomato, cut into wedges
180g Cucumber, cut into chunks
3 pcs Spring onion, cut into
 1-inch pieces
800g Fresh tuna steak
+ Salt, Pepper from the mill

### Salad Nicoise Dressing
100ml Lemon juice
1 clove Garlic
300ml Olive oil
1g Chive, chopped
+ Salt, pepper from the mill

製法 *Procedures*

1 放馬鈴薯在淡鹽水中煮10分鐘至軟身，切成大塊。

2 雞蛋在滾水中煮8分鐘，然後放冷水中冷卻，去殼，切成角形。

3 青豆角焯3分鐘，放入冷水中冷卻。

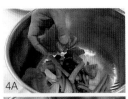
4A

4B

4 馬鈴薯和青菜豆放入碗，加入甜椒、橄欖、番茄、青瓜和青葱，倒入一半沙律醬拌勻。

5 給吞拿魚柳調味，用橄欖油煎至四成熟，取出放碟上待涼5分鐘，切薄片，加上雞蛋和剩餘的沙律醬。

**吞拿魚沙律醬**

1 用刀把蒜頭壓碎。

2 把所有的材料拌在一起，加入調味。

3 取出蒜頭。

1 Boil potatoes in lightly salted water for 10 minutes until tender, cut into chunks.

2 Cook eggs in boiling water for 8 minutes, cool the eggs under cold water, then peel and cut into wedges.

3 Blanch green beans for 3 minutes, cool under cold water.

4 Place potatoes and beans in bowl, add pepper, olives, tomatoes, cucumbers and spring onions, toss the salad with half of the dressing.

5 Well season tuna steak and cook in olive oil until rare, allow to cool for 5 minutes, then slice thinly, arrange on top of salad with egg and remaining dressing.

### Salad Nicosie Dressing

1 Crush the garlic with knife.

2 Whisk all ingredients together, season to taste.

3 Strain garlic from dressing.

*Modern Salad Niçoise*

# 新式吞拿魚沙律

→Tips

選用新鮮吞拿魚柳製作這道特色沙律。

Use fresh tuna for this special salad.

註：傳統吞拿魚沙律常用罐裝吞拿魚製造。

Remarks: Canned tuna is usually used for traditional salad Niçoise.

## 材料 *Ingredients*

| | |
|---|---|
| 蝦36隻（煮熟，去皮，去腸） | 36 pcs Shrimp, cooked, peeled and deveined |
| 冰島萵苣180克（切絲） | 180g Iceberg lettuce, shredded |
| 番茄70克（切丁） | 70g Tomato, diced |
| 帶葉芹菜6條 | 6 pcs Celery rib with leave |
| 檸檬6塊（切楔形） | 6 pcs Lemon wedges |
| 雞尾酒汁180毫升 | 180ml Cocktail sauce |

### 雞尾酒汁     Cocktail Sauces

| | |
|---|---|
| 蛋黃醬250毫升 | 250ml Mayonnaise |
| 番茄醬60毫升 | 60ml Tomato sauce |
| L&P 汁10毫升 | 10ml L & P Sauce |
| 檸檬汁少許 | + Lemon juice |
| 塔瓦斯科辣椒少許 | + Tabasco |
| 鹽、鮮磨胡椒少許 | + Salt, pepper from the mill |

## 製法 *Procedures*

**1** 汁的製法：所有配料混和，以
鹽和胡椒調味。

**2** 蝦去殼去腸。

**3** 萵苣排放到玻璃杯中，輕輕地把蝦
放入，尾巴掛在杯子邊緣。

**1** For the sauce: Mix all ingredients together, season with salt and pepper.

**2** Remove the shells and intestines of the shrimps.

**3** Arrange lettuce in glass, gently put shrimps in with tails outside the glass.

**4** Spoon the sauce in salad and garnish with tomatoes, celery and lemon.

**4** 醬汁舀進沙律，用番茄、芹菜和檸檬裝飾。

*Shrimp Cocktail*

# 鳳尾蝦喀嗲

→ Tips

這道經典的雞尾酒蝦是靠番茄和醬汁提味，而不是辣根來增加辣味，用馬丁利酒杯食用。

This classic shrimp cocktails gets its heat from tomato and sauce instead of horseradish, serve it in martini glass.

| | |
|---|---|
| 橄欖油80毫升 | 80ml Olive oil |
| 茄子600克(切方塊) | 600g Eggplant, cubed |
| 意大利青瓜250克(切片) | 250g Zucchini, sliced |
| 綠甜椒100克(去籽,切丁) | 100g Green bell pepper, seeded and diced |
| 芹菜100克(切方塊) | 100g Celery, cubed |
| 洋葱100克(切片) | 100g Onion, sliced |
| 蒜瓣2個(切碎) | 2 cloves Garlic, chopped |
| 番茄300克(去皮,切角形) | 300g Tomato, peeled, cut into wedges |
| 羅勒菜、俄立岡(意大利香草/牛至葉) | + Basil leaves |
| 百里香葉 | + Oregano leaves |
| 粗海鹽 | + Thyme leaves |
| 鮮磨胡椒少許 | + Coarse sea salt, pepper from the mill |

製法 *Procedures*

**1** 用鍋燒熱油,放入茄子、意大利青瓜、綠甜椒、芹菜、洋葱和蒜頭,煮8分鐘或直至材料軟中帶脆,不時拌炒。

**2** 放入番茄、香料和調味料。

**3** 放上蓋,文火煮至全熟。

**1** Heat oil in pan, add eggplant, zucchini, green bell pepper, celery, onion and garlic, cook for 8 minutes or until crisp-tender, stir occasionally.

**2** Add tomatoes, herbs and season to taste.

**3** Cover, simmer until thoroughly heated.

## *Zesty Ratatouille*
# 法式燴菜

→Tips

這種簡單快速的燴菜法在法國很常用。為了得到最佳的口味,需要使用優質的橄欖油、粗海鹽、鮮磨的黑胡椒和香料。

This quick and easy technique of stewing vegetable is most often used in France. For the best flavour, use good quality olive oil, coarse sea salt, freshly ground black pepper and herbs.

菠菜葉300克(用大量滾水焯熟，用冰水冷卻，瀝乾)

300g Spinach leaves, cooked in lots of boiling water, cool down in ice water and drained.

雞蛋3隻

3 pcs Eggs

高脂厚忌廉250毫升

250ml Double cream

肉豆蔻少許

+ Nutmeg

鹽、鮮磨胡椒少許

+ Salt, pepper from the mill

## 製法 *Procedures*

**1** 用攪拌機打碎菠菜葉、雞蛋、忌廉、肉豆蔻和適量調味料一起攪打成茸。

**2** 把混合物舀進塗了牛油的150毫升烤餡餅模。

**3** 把模子放進熱水盤內，以190℃烤10分鐘或至糕子變硬。

**4** 將模子拿出，用刀弄鬆模邊。

**5** 扣在盤子上，輕輕地拿掉模子。

**1** Purée spinach in blender with eggs, cream, nutmeg and season to taste.

**2** Spoon the mixture into buttered 150 ml timbale moulds.

**3** Put the moulds bain marie in the and bake at 190 ℃ for 10 minutes or until firm.

**4** Remove the moulds from bain marie and run a knife around the inside of moulds to loosen the timbale.

**5** Invert on plate, gently lift off mould.

# *Spinach Timbale*

# 菠菜醬糕

→Tips

1. 如果用小模子來做，做好後倒轉以方便取出。用蔬菜泥如菠菜做，色澤會更誘人，也可以用甘筍或西蘭花。

2. 把一根籤子插到模子，拔出後籤子仍很乾淨，便説明糕已做好了。

1. Puréed vegetables, such as the spinach, will be more attractive when cooked in small moulds and taken out upside-down. Carrot or broccoli can also be used.

2. Insert a skewer into the timbale. When it comes out clean, the timbale is done.

## 材料 _Ingredients_

| | |
|---|---|
| 牛油70毫升 | 70ml Butter |
| 洋蔥90毫升(切碎) | 90ml Onion, finely chopped |
| 長粒米250毫升 | 250ml Long grain rice |
| 熱雞上湯500毫升 | 500ml Hot chicken stock |
| 腰果120毫升(烘香及切粗粒) | 120ml Cashew, roasted and coarsely chopped |
| 番荽60毫升(切碎) | 60ml Parsley, chopped |
| 鹽少許 | + Salt |

## 製法 _Procedures_

**1** 用牛油將洋蔥炒至軟身。

**4** 拌入腰果和番荽。

**2** 加入米飯，拌炒至米粒脫皮，穀粒開始焦黃。

3A

3B

**3** 撒入雞上湯和鹽，蓋上蓋用文火煮15-20分鐘，或待米變軟、湯汁收乾。

**1** Sauté onions in butter until soft.

**2** Add rice, stir until coated and grains start to burst.

**3** Stir in chicken stock and salt, cover and simmer for 15-20 minutes, or until rice is tender and liquid is absorbed.

**4** Stir in cashew and parsley.

*Cashew-rice Pilaf*

# 腰果香飯

→Tips
1. 果仁飯在中東和印度很受歡迎，兩地的廚師都用同樣的方法烹製 — 吸收法。

2. 用作主菜時，可以在最後加入熟肉、海鮮以及蔬菜。

1. Pilaf are popular in the Middle East and India, chefs in both areas make them at the same way - usually by the absorption method.

2. For a main dish, fold in chopped cooked meat or seafood and vegetable before served.

## 材料 *Ingredients*

**6人**
**6 pax**

| | |
|---|---|
| 牛油110克 | 110g Butter |
| 洋蔥100克(切片) | 100g Onion, sliced |
| 蒜頭3瓣(壓碎) | 3 cloves Garlic, crushed |
| 意大利米450克 | 450g Italian risotto rice |
| 乾紅葡萄酒130毫升 | 130ml Dried red wine |
| 牛肝菌(矗士菌)30克 | 30g Cep mushroom |
| 鮮冬菇50克(切片) | 50g Shiitake mushroom, sliced |
| 蘑菇80克(切片) | 80g White mushroom |
| 熱雞上湯1公升 | 1 L Boiling chicken stock |
| 巴馬芝士40克(磨碎) | 40g Parmesan cheese, grated |
| 意大利番荽葉少許(裝飾) | + Italian parsley leaves |
| 鹽、鮮磨胡椒少許 | + Salt and pepper from mill |

## 製法 *Procedures*

**1** 用50克牛油把洋蔥和大蒜炒至軟身。

**2** 撒入米和紅酒加熱，撇去泡沫。

3A

3B

**3** 加入一湯杓煮滾的雞上湯，放於中央，攪拌20分鐘，直至全部湯汁收乾。此時菜飯應該開始變得很柔軟而飽滿、濕潤、有光澤。

**4** 把餘下的上湯一併加入，用細火煮10-15分鐘，不停攪拌。

5A

5B

**5** 用30克牛油把鮮冬菇炒2分鐘，然後與剩下的牛油一起加到湯飯。

**6** 撒入一半芝士和適量調味料拌勻。

**7** 再撒上剩下的芝士，並用意大利番荽作裝飾。

**1** Sauté onion and garlic in 50g butter until soft.

**2** Stir in rice and wine, remove bubbles.

**3** Add a ladle of boiling stock in the middle, stir for about 20 minutes until all the liquid is used and absorbed. The risotto should be tender but still very rich, moist and glossy.

**4** Add all the stock at once and cook over low heat for 10-15 minutes, stir constantly.

**5** Sauté the mushroom in 30g butter for 2 minutes, then add into risotto with remaining butter.

**6** Stir in half of the cheese and season to taste.

**7** Sprinkle with the remaining cheese and garnish with parsley.

# *Mushroom Risotto*
# 白菌意大利飯

## →Tips

1. 這道風行的意大利飯的特色是加入了牛肝菌，鮮冬菇和蘑菇。

2. 意大利飯要有奶狀效果，可用短米如意大利 Arborio 或 Carnaroli 米。

1. Here, the popular Italian rice dish risotto is given an elegant finish with Cep mushroom, Shiitake mushroom and White mushroom.

2. To make it creamy, be sure to use short grain rice such as Arborio or Carnaroli.

## 材料 *Ingredients*

6人
**6 pax**

|  |  |
|------|---------|
| 魷魚6隻(修剪) | 6 pcs Squid, trimmed |
| 黑青口600克(擦淨) | 600g Black mussel, scrubbed |
| 白葡萄酒120毫升 | 120ml White wine |
| 橄欖油80毫升 | 80ml Olive Oil |
| 雞肉400克(切大塊) | 400g Chicken meat, cut into chunks |
| 西班牙辣香腸400克(切大塊) | 400g Chorizo, cut into chunks |
| 中蝦6隻(去頭,留尾) | 6 pcs Prawns without tails, heads |
| 淡水螯蝦或海螯蝦6隻 | 6 pcs Yabbies or langoustines |
| 洋蔥120克(切厚片) | 120g Onion, thickly sliced |
| 紅甜椒100克(切粒) | 100g Red bell pepper, thickly sliced |
| 蒜頭1瓣(切碎) | 1 clove Garlic, chopped |
| 番茄100克 | 100g Tomato, chopped |
| 紅椒粉(甜椒粉)10毫升 | 10ml Paprika |
| 魚湯1.2公升 | 1.2 L Fish stock |
| 意大利米400克 | 400g Risotto rice |
| 蠶豆180克(去皮) | 180g Broad beans, skinned |
| 意大利番荽30毫升(切粗粒,裝飾) | 30ml Italian parsley leaf, roughly chopped |
| 藏紅花少許 | + Little Saffron |

## 製法 *Procedures*

**1** 用白酒煮青口至開口,上蓋,保留湯汁待用。

**2** 用橄欖油把中蝦、螯蝦和魷魚炒1分鐘,待用。

**3** 用同一隻鍋炒雞肉和香腸,炒6分鐘至啡色。

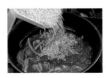

**5** 加入米,不用上蓋,用文火煮12-14分鐘或直至米飯幾乎全熟,經常攪拌。

**6** 加入蠶豆、藏紅花,剩餘的紅椒粉和少許油,拌勻。

**7** 加入所有海鮮,以慢火煮6-8分鐘,直至米飯全熟收乾,如有需要可多加些上湯。

**8** 撒上番荽奉客。

**4** 加入洋蔥、紅甜椒、蒜頭、番茄、上湯、葡萄酒、湯汁、煮青口白酒湯汁和5毫升紅椒粉,煮至滾。

**1** Boil mussels in white wine, covered, set aside and keep the liquid.

**2** Sauté prawns, yabbies and squid in olive oil for 1 minute, set aside.

**3** Use the same pan to cook chicken and chorizo, brown them for 6 minutes.

**4** Add onion, bell pepper, garlic, tomatoes, stock, wine liquid and 5 ml of paprika, bring to the boil.

**5** Stir in rice, uncovered, simmer and stir for 12-14 minutes or until rice almost cooked.

**6** Add beans, saffron, remaining paprika and some extra oil, mix well.

**7** Add all seafood and cook over very low heat for 6-8 minutes until rice is fully cooked and dry, add extra stock if necessary.

**8** Sprinkle parsley on top and serve.

# *Spanish Paella with Chicken, Prawn and Squid*

# 西班牙海鮮飯

## →Tips

1. 西班牙海鮮飯顏色鮮艷，只用一隻鍋便能做好。可以用短顆粒的西班牙paella米來做，如Bomba，也可以用意大利risotto一類的米。

2. 西班牙海鮮飯烹煮時較少攪拌，米飯直接倒進湯汁，與意大利的risotto飯有很大分別。

1. Spanish paella is a colourful, one-pan dish. Use short- grain Spanish paella rice such as Bomba, or an Italian risotto style rice.

2. Less stirring and rice is directly put into the liquid make this dish very different from Risotto.

## 材料
### Ingredients

| | |
|---|---|
| 清水900毫升 | 900ml Water |
| 玉米粉150克 | 150g Polenta |
| 牛油40克（室溫） | 40g Butter (room temp.) |
| 巴馬芝士20克（磨碎） | 20g Parmesan cheese, grated |
| 鹽少許 | + Salt |

## 製法 Procedures

**1** 鹽加入水中煮滾，轉用慢火保持溫度。

**2** 徐徐加入玉米粉，不停攪拌，煮約15分鐘直至湯水不再黏稠。

**3** 放入牛油和芝士，拌勻即成。

### 煎玉米糕

**1** 趁熱把玉米糕攤扒在器皿上，約開2厘米厚，放涼。

**2** 把玉米糕切開，掃點橄欖油。

**3** 用平底鍋煎或炭烤至呈金黃色。

**1** Bring water and salt to boil, reduce heat to simmer.

**2** Slowly add polenta, stir constantly, cook for about 15 minutes until it is not sticky.

**3** Stir in butter, cheese and serve .

**For fried polenta**

**1** Omit butter and cheese when polenta is still hot. Spread it 2-cm thick on clean surface, let cool.

**2** Cut the polenta and brush with olive oil.

**3** Pan-fry or charbroil until golden brown.

# *Polenta*
# 玉米糕

→Tips

Polenta 也叫作cornmeal，可以作為粥來食用，可加入牛油和巴馬芝士，成為小菜，也可以用鍋煎或炭烤方法製成酥脆的玉米菜式。玉米粒是意大利北部地區日常飲食的一部份，通常與燉肉和醬汁一起食用。

Polenta is also called ornmeal it can be served moist, enriched with butter and Parmesan cheese as a side dish, or in firm, crisp piece by pan-frying or charbroiling. Polenta is a part of a diet in northern Italy, usually served with stew and sauce meat dish.

## 材料 Ingredients

| 牛油50克 | 50g Butter |
| 煙肉150克(切碎) | 150g Sacked beacon, chopped |
| 新鮮迷迭香葉5毫升(切碎) | 5ml Fresh rosemary leaves, chopped |
| 洋葱120克(切細) | 120g Onion, finely chopped |
| 蒜頭2瓣(切細) | 2 cloves Garlic, finely chopped |
| 甘筍100克(切細) | 100g Carrot, finely chopped |
| 芹菜80克(切細) | 80g Celery, finely chopped |
| 免治瘦牛肉450克 | 450g Lean minced beef |
| 乾紅葡萄酒100毫升 | 100ml Dry red wine |
| 番茄茸250毫升 | 250ml Tomato purée |
| 曬乾番茄50毫升(切細) | 50ml Sun-dried tomato, finely chopped |
| 牛肉上湯200毫升 | 200ml Beef stock |
| 乾俄立岡(乾牛至)5毫升 | 5ml Dried oregano |
| 意大利粉500克 | 500g Dried spaghetti |
| 新鮮羅勒少許 | + Fresh basil |
| 巴馬芝士少許(磨碎) | + Parmesan cheese, grated |
| 鹽、鮮磨胡椒少許 | + Salt, pepper from the mill |

## 製法 Procedures

**1** 用一半牛油將煙肉和迷迭香炒至微呈金黃色，然後加入洋葱、蒜頭、甘筍、芹菜，炒3-4分鐘至蔬菜軟身。

**2** 加入剩下的牛油，再加入牛肉，拌炒2-3分鐘。

**3** 倒入葡萄酒，煮至湯汁微微變稠，然後放入全部番茄、牛肉上湯和俄立岡，放入調味。煮滾，用慢火炆煮，加上蓋，用慢火再煮2小時。

**4** 在一大鍋鹽水中煮意粉，要煮得不軟，仍有嚼勁，瀝乾，把醬汁置於碟上，放上意粉，再撒上羅勒葉和芝士。

**1** Sauté bacon and rosemary with half of the butter until lightly golden, then add onion, garlic, carrot, celery and fry for 3-4 minutes until softened.

**2** Add remaining butter and then beef, stir and fry for 2-3 minutes.

**3** Add wine and reduce slightly, then add all tomatoes, beef stock and oregano, season to taste. Bring to the boil and simmer, covered, cook over very low heat for 2 hours.

**4** Cook the pasta in a large pot of boiling salted water until al dente, drain. Place the sauce on plate and arrange the pasta on top then add basil and sprinkle with the cheese.

# *Spaghetti Bolognese*
# 肉醬意大利粉

→Tips

1. 傳統做法，肉醬是配 tagliatelle 麵條食用的，在這裏我們改配了 spaghetti。

2. 意粉要煮到仍然有嚼勁，意大利人稱為 al dente。如果煮過熟，會變成糊。

1. Traditionally, Bolognese was served with tagliatelle, but now we serve it with spaghetti.

2. Pasta should be cooked until al dente which is called by the Italians and meant firm to bite. If it is overcooked, it will be mushy.

## 材料
## *Ingredients*

**6 pax**

| | |
|---|---|
| 乾扁麵500克 | 500g Dried Linguine |

| 香草醬料 | **For the Pesto** |
|---|---|
| 新鮮羅勒葉80克 | 80g Fresh basil leaves |
| 新鮮番荽葉20克 | 20g Fresh Parsley leaves |
| 蒜頭3瓣（切碎） | 3 cloves Garlic |
| 松子仁50克（烘乾） | 50g Pine nut, roasted |
| 巴馬芝士100毫升 | 100ml Parmesan cheese, grilled |
| 橄欖油130毫升 | 130ml Olive oil |
| 鹽少許 | + Salt |

## 製法 *Procedures*

1A    1B

**1** 把羅勒、西芹、蒜頭、果仁、芝士和鹽與一半油一起放入攪拌機內打碎。

2A    2B

**2** 徐徐加入剩下的油，直至混和物成糊狀。

### 煮扁麵

**1** 在一大鍋煮滾的鹽水中煮麵條，煮至麵條仍有咬勁，瀝乾。

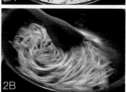

2A

2B

**2** 把香草醬和麵條拌在一起，直至醬料均勻地沾在麵條上。

**1** Work basil, parsley, garlic, nut, cheese and salt with half the oil in food processor until finely chopped.

**2** Slowly add the remaining oil until a paste is formed.

### Cooking pasta

**1** Cook the pasta in large pot of boiling salted water until al dente, drain.

**2** Toss the sauce with warm pasta until pasta is well-coated.

# *Linguine with Pesto Sauce*

# 香草醬扁麵

→Tips

1. 傳統做法是用香草醬配 linguine 扁麵，但你也可選配任何麵條進食。

2. 香草醬可在一星期前做好，放入密封的容器冷藏。

3. 在醬上面倒一些油，可以防止醬色變黑。

1. Traditionally, linguine is served with pesto but you can go with any pasta you like.

2. Pesto sauce can be made up to one week in advance and refrigerated in an airtight container.

3. Pour some extra oil over top of sauce to prevent pesto turns black.

| | |
|---|---|
| 乾螺絲粉500克 | 500g Dried fusilli |
| 煙肉10片 | 10 Slices of Smoky bacon |
| 橄欖油5毫升 | 5ml Olive oil |
| 蛋黃5隻 | 5 pcs Egg yolk |
| 高脂厚忌廉100毫升 | 100ml Double Cream |
| 巴馬芝士125克（搓碎） | 125g Parmesan cheese, grated |
| 鹽、鮮磨胡椒少許 | + Salt, pepper from the mill |

製法 *Procedures*

**3** 麵煮熟時，瀝乾，馬上與煙肉和蛋黃液拌在一起，直至醬汁均勻地塗在麵條上。

**4** 用慢火煮麵至醬汁略微變稠，加入適量調味。

**1** 把麵條放進煮滾的鹽水中煮至有咬勁，一邊煮麵條；另一邊把煙肉放油裏煎至脆，略微搗碎，待用。

**2** 把蛋黃、忌廉和一半芝士放在碗裏打勻。

**1** Cook pasta in salted boiling water until al dente. When cooking pasta, slowly fry bacon in oil until crispy, break it up a bit, set aside.

**2** Whip up egg yolk, cream and half of the cheese in a bowl.

**3** When pasta is cooked, drain and immediately toss it with bacon and egg mixture until the sauce is coated with pasta.

**4** Cook the pasta in very low heat until the sauce slightly thickened, season to taste.

*Fusilli Carbonara*

# 煙肉蛋黃汁螺絲粉

→Tips

1. 做烤麵條的概念是用蛋黃來使醬汁變稠，並取煙肉和忌廉汁的味道。

2. 不要煮過熟，否則會變一團蛋黃糊。

1. The concept of making Carbonara is to use egg to thicken the sauce and get the flavour from bacon and cream sauce.

2. Do not over cook otherwise it will become a scrabbled egg sauce.

## 材料
## *Ingredients*

| | |
|---|---|
| 新鮮三文魚柳2片，每片900克(連皮) | 900g x 2 Fresh Salmon fillet, unskinned |
| 海鹽75克 | 75g Sea salt |
| 糖125克 | 125g Sugar |
| 白胡椒粒10毫升 | 10ml White peppercorn |
| 新鮮蒔蘿(刁草)(切粗)100毫升 | 100ml Fresh dill, coarsely chopped |
| 芥末和蒔蘿汁125毫升 | 125ml Mustard and dill sauce |
| 檸檬角3片 | 3 pcs Lemon wedge |

### 芥末蒔蘿(刁草)汁料
### Mustard and Dill Sauce

| | |
|---|---|
| 蛋黃醬125毫升 | 125ml Mayonnaise |
| Pommery 芥末15毫升 | 15ml Pommery mustard |
| 蒔蘿(刁草)籽3毫升 | 3ml Dill weed |

## 製法 *Procedures*

**1** 用刀把胡椒粒壓碎。

**2** 把三文魚柳放在淺盤中，有皮的一邊朝下。

**3** 把海鹽、糖和胡椒粒混和，撒在魚柳上。

**4** 把香草均勻地撒在鹽上。

**5** 把兩片魚柳皮朝上疊在一起，把錫紙包住的薄皮壓在魚柳上。放入冰箱冷藏3天，每12小時翻動一次，直到調味料滲透魚肉中。

**6** 把魚柳分開，切成薄片，排放在盤子上，配檸檬、蒔蘿以及芥末蒔蘿汁進食。

**7** 蒔蘿醬汁：所有配料混和拌勻。

**1** Crush peppercorn with knife.

**2** Put salmon fillets in a shallow dish with skin side down.

**3** Combine sea salt, sugar and peppercorn and sprinkle over the fish.

**4** Sprinkle herb evenly over the salt mixture.

**5** Lay the uncoated fillet, skin side up over the other. Place foil-covered cardboard over the filets and press them. Refrigerate for 3 days, turn every 12 hours until the seasoning have penetrated the flesh.

**6** Separate the two fillets and cut into thin slices, put the slices on plate and serve with lemon and dill, and mustard dill sauce.

**7** Mustard and Dill Sauce: Combine all ingredients together, mix well.

# Gravadlax with Mustard Dill Sauce

# 香草醃三文魚

→Tips

在瑞典，人們以完美的醃魚技法，製作了著名的gravadlax。醃製未去皮的三文魚柳，然後存放於冰箱2天。

In Sweden, people have perfect art of salting fish to produce the famous gravadlax. Just use unskinned salmon fillet, once marinated, store in the refrigerator for up to 2 days.

## 材料 *Ingredients*

| | |
|---|---|
| 鯖魚柳370克(不去皮) | 370g Mackerel fillet, unskinned |
| 鯛魚柳370克(不去皮) | 370g Snapper fillet, unskinned |
| 龍脷柳370克(不去皮) | 370g Sole fillet, unskinned |
| 魚上湯1公升 | 1 L Fish stock |
| 青瓜香草汁180毫升 | 180ml Cucumber and herb sauce |
| 鹽、鮮磨胡椒少許 | + Salt, pepper from the mill |

### 青瓜和香草汁料　　*Cucumber and Herb Sauce*

| | |
|---|---|
| 酸忌廉125毫升 | 125ml Sour Cream |
| 進口青瓜100克 | 100g Imported cucumber |
| 番荽15毫升(切碎) | 15ml Parsley, chopped |
| 蒔蘿(刁草)5毫升(切碎) | 5ml Dill, chopped |
| 檸檬汁5毫升 | 5ml Lemon juice |
| 鹽、鮮磨胡椒少許 | + Salt, pepper from the mill |

## 製法 *Procedures*

1A

1B

**1** 將魚柳切成20厘米×2厘米的魚片，每三條魚片一組，魚皮朝上，編成辮。

2A

2B

**2** 給魚柳調味，放入微滾的魚上湯 蒸大約8-10分鐘。

**3** 配青瓜香草汁食用。

4A

4B

**4** 青瓜香草汁：青瓜去皮切細；將所有材料拌在一起。

**1** Cut each fillet into strips for about 20 x 2cm, lay three strips with skin side up, interweave the strips.

**2** Season the fish and steam the plaits in simmering fish stock for about 8-10 minutes.

**3** Serve with cucumber and herb sauce.

**4** Cucumber and Herb Sauce: Peel and finely chop the cucumber, combine all ingredients together.

# *Steaming Fish Plaits*

# 清蒸魚辦

→Tips

1. 用不同形狀的魚柳和不同顏色的魚肉，如鯖魚，鯛魚和龍脷魚，可以令這道菜賣相更吸引。

2. 清蒸是保存魚肉質感的最佳烹調方法。

1. A variety of round and flat fish fillets with different flesh and colors, such as mackerel, snapper and sole, can be used to make a more impressive presentation.

2. Steaming is the best cooking method to preserve the texture of the fish.

## 材料 *Ingredients*

2人
**2 pax**

| | |
|---|---|
| 白魚（原條彩虹鮭魚）400克 | 400g White fish (Whole Rainbow Trout) |
| 青檸2個 | 2 pcs Lime |
| 新鮮迷迭香少許 | + Fresh rosemary |
| 鹽，鮮磨胡椒少許 | + Salt, pepper from the mill |
| 橄欖油少許 | + Olive oil |

## 製法 *Procedures*

**1** 青檸一開為二。

**2** 把魚放在塗了油的烤架上，撒上香草和調味料，收緊魚烤架，燒烤魚的兩面3分鐘，不時塗刷橄欖油。

**1** Cut lime into halves.

**2** Place the fish in oiled rack with herb and seasonings, close the fish rack tightly, cook the fish on both sides and griddle for 3 minutes, baste frequently with olive oil.

**3** Check for doneness: The skin should be crisp and golden and the fork inserted is clean when coming out.

**4** Grill the lime and serve with fish.

**3** 檢查是否燒熟：皮變脆，呈金黃色，把叉子叉進去，拔出來的叉子未沾魚肉。

**4** 烤一下青檸，配上魚一起食用。

# *Barbecuing Whole Fish*
# 炭烤全魚

→Tips

1. 燒烤的熱力能令魚燒乾和濃縮魚味。這個菜譜用的是鮭魚，也可用其他魚如鯖魚、鱸魚或吞拿魚取代。

2. 要把全魚均勻烤熟，可在魚身先劃上長條切口；最好是用專門的魚烤架，便於操作。

1. The heat of the barbecue makes the fish dry and tasty. Trout is used in this recipe but other suitable fish such as mackerel, bass or tuna can also be used.

2. For even cooking, score the fish and for the ease of handling, it'll be better to use a fish rack.

## 材料 *Ingredients*

6人
**6 pax**

| | |
|---|---|
| 鱈魚柳6塊（每塊140克） | 6 Seabass fillet (140g each) |
| 芫荽4棵 | 4 sprigs Coriander |
| 甘筍180克（切絲） | 180g Carrot, julienne |
| 乾白葡萄酒60毫升 | 60ml Dry white wine |
| 鹽、鮮磨胡椒少許 | + Salt, pepper from the mill |

## 製法 *Procedures*

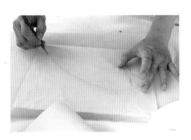

**1** 用防油紙剪出一個心形，應比魚柳每邊長5厘米，抹上油。

2A

2B

**2** 把已調味的魚柳與芫荽、甘筍和白葡萄酒一起放在心形的一邊。

3A

3B

**3** 把紙的另一半合起來，沿邊扭起，封住邊緣。

**1** Greaseproof paper: Cut a heart shape 5 cm larger than the fish out of grease-proof paper, oil it.

**2** Put the seasoned fish on one side of the heart with coriander, carrot and wine.

**3** Fold over the other half of the paper and twist to seal the edges.

**4** Place on baking sheet and bake at 180℃ for 15-20 minutes until puffed.

**4** 放在烤盤上，以180℃烤15-20分鐘，直至紙包膨脹起來。

# *Baking en Papillote*

# 紙包焗魚

→ Tips

en papillote 是個法語詞,意為「紙包」,特點是保護魚肉濕潤不變乾。配料可用香草、蔬菜和酒,在烹調時可以增添菜餚的口味。

The term en papillote is French which means "in a paper bag". It helps to protect the fish and keep it moist. The topping, mainly the herbs, vegetables and white wine adds flavour during cooking.

魚柳12塊(每塊60克)
裹魚用的麵粉100毫升
鹽,鮮磨胡椒少許

60g x 12 Fish fillet
100ml Flour for coating the fish
+ Salt, pepper from the mill

麵糊料
麵粉225克(已篩),發粉8毫升
雞蛋2隻(打勻),牛油(溶化)20毫升
溫水150毫升,啤酒150毫升
鹽少許

**For the batter**
225g Plain flour, sifted
8ml Baking powder, 2 pcs Eggs, beaten
20ml Butter, melted, 150ml Warm water
150ml Beer,  Salt

他他汁料
雞蛋1隻(煮熟,切碎)
酸瓜1條(切碎)
水瓜鈕(續隨子)5毫升(切碎)
乾2克(切碎)
番荽2.5毫升(切碎)
他力芹香草(茵陳高)2.5毫升(切碎)
蛋黃醬120毫升
鹽、鮮磨胡椒少許

**Tartar Sauce**
1 pc Egg, cooked & chopped
1 pc Pickle, chopped
5ml Capers, chopped
2g Shallot, chopped
2.5ml Parsley leaves, chopped
2.5ml Tarragon leaves, chopped
120ml Mayonnaise
+ Salt, pepper from the mill

製法 Procedures

**1** 把麵粉、發粉和鹽混
和,待用。

**3** 用鹽和鮮磨胡椒給魚
輕輕調味,裹上麵粉。

5A

5B

**5** 他他汁:把所有配料
混和,適量調味。

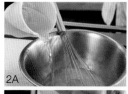
2A

2B

**2** 攪打雞蛋、牛油、水
和啤酒,加入麵粉,
打至柔滑。

**4** 把魚放到麵糊中蘸一
下,瀝去多餘麵糊。
放在180℃的熱油中
炸4-5分鐘至金黃色
及可用叉輕易分開魚
柳為合。用紙巾瀝乾
油,配他他汁食用。

**1** Mix the flour, baking powder and salt together, set aside.

**2** Beat eggs, butter, water and beer, add flour, beat until smooth.

**3** Season the fish with salt and pepper lightly, coat with flour.

**4** Dip fish in batter, drain off the excess. Fry at 180℃ hot oil for 4-5 minutes or until golden brown and fish can be separated easily by fork. Drain with paper towel and serve with tartar sauce.

**5** Tartar sauce: mix all ingredients together, season to taste.

*Deep-frying*
*Fish in Batter*

# 脆炸魚柳

→Tips

麵糊可以形成一個保護膜，讓魚肉保持濕潤，用高溫烹調和使用植物油可以得到最好的效果。

Batter provides a protective coating which keeps fish moist. High cooking temperature (180℃ ) is necessary for the best result, and vegetable oil are best to use.

| | |
|---|---|
| 三文魚柳6片（每片120克） | 120g x 6  Salmon fillet |
| Cajun 調味料20克 | 20g Cajun seasoning |
| 牛油清60毫升 | 60ml Clarified butter |
| 檸檬6個 | 6 pcs Lemon |
| Mesclun 沙律菜120克 | 120g Mesclun salad vegetable |
| 油醋汁30毫升 | 30ml Vinaigrette |

## 製法 *Procedures*

**1** 三文魚塗上Cajun調味料調味。

4A 4B

**4** 烤魚、沙律和檸檬配合食用。

2A 2B

**2** 魚放入熱牛油中，煎約3分鐘，直至外層變微焦，其間翻動一次。

**1** Coat salmon with Cajun seasonings.

**2** Place fish in hot butter, fry for about 3 minutes until the coating is charred, turn once.

**3** Toss the salad with dressing.

**4** Serve the fish with salad and lemon.

**3** 沙律菜加上油醋汁。

# *Broiling Cajun-style Salmon Steak*

# 美式煎三文魚

→Tips

Cajun 式煎魚源自新奧爾良和墨西哥，外面是一層深色、熱烘烘的胡椒脆皮，這層特殊的醃料叫 Cajun 調味料，有許多種配方。我經常採用 Luzianne 配方，可配肉類使用。

Fish fried in Cajun-style comes from New Orleans and Mexico and has a dark, pepper hot crust, this is a result of a special coating called Cajun seasoning which has many recipes, I always use the Luzianne one, it is good for meat too.

## 材料
### Ingredients

ᗖ6人
**6 pax**

| | |
|---|---|
| 龍蝦3隻 | 3 pcs Lobster |
| 橄欖油30毫升 | 30ml Olive oil |
| 洋葱150克(切碎) | 150g Onion, finely chopped |
| 乾葱30克(切碎) | 30g Shallot, finely chopped |
| 牛油60克、麵粉30克 | 60g Butter, 30g Flour |
| 牛奶400毫升 | 400ml Milk |
| 細葉芹6毫升(切碎) | 6ml Chervil, chopped |
| 他力芹香草(茵陳蒿)4毫升(切碎) | 4ml Tarragon, chopped |
| 番荽6毫升(切碎) | 6ml Parsley, chopped |
| 法式芥末10毫升 | 10ml Dijon mustard |
| 白葡萄酒120毫升 | 120ml White wine |
| 高脂厚忌廉60毫升 | 60ml Double cream |
| 巴馬芝士45毫升(磨碎) | 45ml Parmesan cheese, grated |
| 荷蘭汁60毫升 | 60ml Hollandaise sauce |
| 鹽、鮮磨胡椒少許 | + Salt, pepper from the mill |

## 製法 Procedures

**1** 用長竹籤把龍蝦的尿液放出。

 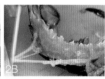

**2** 切開龍蝦，去腸，將龍蝦肉從殼 取出，待用。

**3** 在龍蝦殼上滴幾滴橄欖油，敲開龍蝦的鉗子，取出龍蝦肉。

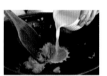

**4** 爆香洋葱，加入牛油炒至軟身，不要炒至變色。

**5** 撒入麵粉煮3分鐘，不停拌炒。

**6** 離火徐徐倒入牛奶拌勻，回火煮滾，以慢火煮至黏稠得宜。

**7** 加入香草、芥末和調味料。離火，拌入酒和忌廉。用慢火煮5分鐘，直至湯汁變回黏稠。

**8** 放入龍蝦肉再以慢火煮2-3分鐘（不要把龍蝦肉煮過熟），伴入荷蘭汁。

**9** 把肉汁舀到龍蝦殼上，撒上芝士，放烤架上烤至呈棕黃色，用番荽裝飾。

**1** Release urine from lobster by wooden stick.

**2** Split the lobster, remove the vein, take the meat from the shell and set aside.

**3** Polish the shell with few drops of olive oil, crack the claw and remove the meat from them.

**4** Saut onion and add in butter until tender, do not brown.

**5** Stir in the flour and cook for 3 minutes, stir all the time.

**6** Remove from heat, gradually blend in the milk, bring to the boil and simmer to the right consistency.

**7** Add herbs, mustard and seasonings; remove from heat and whisk wine and cream, return to low heat and simmer for 5 minutes until return to high consistency again.

**8** Add lobster meat to the sauce and simmer for 2-3 minutes (do not over-cook the lobster), mix with Hollandaise sauce.

**9** Spoon the mixture into the lobster shell, top with cheese and brown under the grill. Garnish with parsley.

*Lobster Thermidor*

# 龍蝦"米多"

→Tips

龍蝦肉多、清甜、肉質細膩。要吃到鮮甜的龍蝦，最好是買活的。

Lobster flesh is meaty, sweet and delicate. For the best result, buy lobsters when they are still alive.

## 材料 Ingredients

活青口1公斤（洗淨）　　1kg Fresh mussels, cleaned
牛油35克，粗鹽600克　　35g Butter, 600g Rock salt
乾葱30克（切碎）　　　　30g Shallot, chopped
蒜頭30克（切碎）　　　　30g Garlic, chopped
乾身白酒250毫升　　　　250ml Dry white wine
番荽20毫升（切碎）　　　20ml Parsley, chopped
蒜茸香草牛油100克　　　100g Garlic herb butter
鹽、鮮磨胡椒少許　　　　+ Salt, pepper from the mill

### 蒜茸香草牛油料 / Garlic Herb Butter

無鹽軟身牛油120克　　　120g Softened unsalted butter
百里香葉1克（切碎）　　　1g Thyme leaves, chopped
番荽葉4克（切碎）　　　　4g Parsley leaves, chopped
蒜頭2瓣（切碎）　　　　　2 cloves Garlic, chopped
鹽、鮮磨胡椒少許　　　　+ Salt, pepper from the mill

## 製法 Procedures

**1** 放牛油在深鍋中，炒乾葱和蒜茸至軟身。

**2** 放入酒和番荽，以慢火炆煮。

**3** 放入青口，加蓋蒸6-7分鐘至青口打開，把未開殼的青口撿出不要，然後把可用青口的半邊殼撕去。

**4** 在烤盤裏把青口排放在粗鹽上。

**5** 在每隻青口上放上蒜茸香草牛油，烤焗2-3分鐘。

### 蒜茸香草牛油：

**1** 攪打牛油2-3分鐘，直至牛油變透明，呈軟狀，把所有材料拌在一起。

**2** 把混合物放在保鮮紙上，捲起來，做成條狀，放於冰箱內冷藏6小時。

**1** Sauté the shallot and garlic in a deep pan with butter for 5 minutes until soft.

**2** Add wine and parsley, bring to simmer.

**3** Add in mussels, cover and steam for 6-7 minutes until the mussels open, discard those remain closed, then open them with fingers and discard the top shells.

**4** Arrange the needed shells on rock salt in a baking dish.

**5** Top each mussel with garlic herb butter and bake for 2-3 minutes.

### Garlic Herb Butter:

**1** Beat butter for 2-3 minutes until light and creamy, mix all ingredients together.

**2** Pass the butter mixture into plastic wrap, roll it up and keep in refrigerator for 6 hours.

# *Mussels with Garlic Herb Butter*
# 香草蒜茸牛油焗青口

### ✦Tips

1. 選用完整而帶鮮味的青口。

2. 不要挑選重身的青口,面可能滿佈沙石;也不要選搖起來很輕很鬆的青口(可能是死的)。一定要選貝殼緊閉的青口。

1. Choose undamaged, fresh smelling mussels.

2. Avoid mussels which are heavy  they may be full of sand, or light and loose when shaken (they are probably dead). Ensure all are tightly closed.

| | |
|---|---|
| 雞1公斤（修剪洗淨） | 1kg Chicken, trimmed and cleaned |
| 洋蔥½個 | ½ pc Onion |
| 牛油60克（切片） | 60g Butter, sliced |
| 煙肉2塊（切片） | 2 pcs Bacon, sliced |
| 麵粉15毫升 | 15ml Plain flour |
| 熱雞上湯500毫升 | 500ml Hot chicken stock |
| 鹽、鮮磨胡椒少許 | + Salt, pepper from the mill |

## 製法 *Procedures*

**1** 用紙巾把雞擦乾，雞肚塗上調味料，再放入半個洋蔥。

 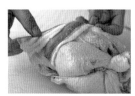

**2** 雞胸朝上，把繩子繫在雞腿上和雞屁股下方。

**5** 把雞放在烤盤的架子上，用煙肉蓋住雞胸，放入200℃的焗爐中烤。

**3** 繩子通過雞腿和雞身拉到頸子處。

**6** 雞身的一側焗15分鐘後，再翻過另一側焗15分鐘，然後把雞胸朝上，再焗20分鐘。

**7** 把雞從烤盤中取出，倒掉油脂，只留30毫升汁液，撒入麵粉拌勻。

**4** 把雞翻過身來，用繩子把兩隻翅膀拴在一起，然後打一個紮實的結。

**8** 邊攪邊把雞湯加入並煮滾，以慢火煮2分鐘，放入適量調味料，與切片的雞肉一起食用。

**1** Wipe the bird both inside and outside with kitchen paper. Season the cavity and insert half an onion.

**2** With the chicken breast side up, tie string around the legs and under the skin flap at the tail.

**3** Bring the string towards the neck of the chicken, pass it down between the legs and body.

**4** Turn the chicken over, pull the string to bring the wings together, then tie a firm knot.

**5** Put the chicken on a rack in tin, cover the breast with bacon, roast the chicken in an 200℃ oven.

**6** Start cooking the chicken on one side for 15 minutes, then turn it over and cook on the other side for 15 minutes, turn the breast side up and roast for another 20 minutes.

**7** Remove the chicken from tin and pour out the fat, just keep about 30 ml, sprinkle in flour and stir well.

**8** Gradually whisk hot stock, bring to the boil, simmer and whisk for 2 minutes, season to taste, serve sauce with sliced chicken.

# Roasting a
# Whole Chicken
# 烤全雞

→Tips

家禽和野禽肉的脂肪很少，所以在燒烤前要把油脂塗在外面，在燒烤時令肉質保持濕潤。

Poultry and game birds have little natural fat, so to ensure the meat stay moist during roasting, place fat on the skin before roasting.

## 材料 *Ingredients*

| | |
|---|---|
| 雞 1.2 公斤 (去骨) | 1.2kg Chicken, boned |
| 雞上湯 2 公升 | 2 L Chicken stock |
| 鹽、鮮磨胡椒少許 | + Salt, pepper from the mill |

**餡料**

雞胸 350 克
(去骨,去皮,切成塊)
新鮮麵包糠 60 克
牛奶 30 毫升
牛油 15 毫升
乾葱 1 粒 (切碎)
蒜頭 1 瓣 (切碎)
蛋白 1 隻
雜香草 15 克 (百里香、他力芹香草或鼠尾草) (切碎)

**Stuffing**

350g Chicken breast, boneless, skinless, cut into pieces
60g Fresh breadcrumbs
30ml Milk
15ml Butter
1 pc Shallot, chopped
1 clove Garlic, chopped
1 pc Egg white
15g Mixed herbs (Thyme, Tarragon or Sage), chopped

## 製法 *Procedures*

**1** 餡料:用攪肉機攪碎雞肉。

**2** 麵包糠泡在牛奶 至完全浸透,倒出多餘的牛奶,把麵包糠加到攪碎的雞肉 。

**3** 用牛油炒乾葱和蒜頭至軟身,放涼,加入碎雞肉中,拌匀。

**4** 雞肉和蛋白拌匀,放入香草,加入適量調味料。

**5** 於雞肉內調味,接着把餡料均匀地抹在着面,然後把兩邊的雞肉拉起來蓋住餡料。

**6** 在雞的表面擦上鹽和鮮磨胡椒以調味。

7A

7B

**7** 用油紙把雞捲起來做成筒形,然後用錫紙包起來,用繩子紮好。

**8** 把包好的雞放入上湯中以慢火炆 30 分鐘。

**9** 雞肉卷切片,雪凍後以番茄花作裝飾奉客。

**1** Stuffing: Mince the chicken meat in food processor.

**2** Soak breadcrumbs in milk until absorbed, then squeeze out the excess milk and add bread to the minced chicken.

**3** Sauté shallot and garlic in butter until softened,
let cool then add to the chicken mixture, mix well.

**4** Blend the mixture with egg white, then add herbs and season to taste.

**5** Season the inside of the whole boned chicken. Spread the inside evenly with stuffing and pull up the side of the chicken to cover the stuffing.

**6** Season the outside of the chicken by rubbing the skin with salt and pepper.

**7** Roll the chicken with grease paper to make a cylinder, then wrap with aluminum foil, tie the roll with string .

**8** Poach the chicken roll in stock for 30 minutes.

**9** Slice the chicken roll and serve cold, garnish with tomato rose.

*Stuffed Chicken Roll*

# 釀雞卷

→Tips

用上湯炆雞肉卷是非常美味的一道菜。

One of the most delicate dish is to poach chicken roll in flavoured broth.

## 材料 *Ingredients*

| | |
|---|---|
| 雞1隻(1.2公斤)(切成12塊) | 1 pc (1.2kg) Chicken, cut into 12 pieces |
| 菜油60毫升 | 60ml Vegetable oil |
| 麵粉15毫升 | 15ml Plain flour |
| 甘筍250克(切片) | 250g Carrot, sliced |
| 洋蔥1個(切碎) | 1 pc Onion, chopped |
| 月桂葉2片 | 2 pcs Bay leave |
| 蘑菇170克(切片) | 170g Mushroom, sliced |
| 鹽、鮮磨胡椒少許 | + Salt, pepper from the mill |

### 醃料 / Marinade

| | |
|---|---|
| 紅葡萄酒700毫升 | 700ml Red wine |
| 甘筍1個(切碎) | 1 pc Carrot, chopped |
| 洋蔥1個(切碎) | 1 pc Onion, chopped |
| 蒜頭2瓣(切碎) | 2 cloves Garlic, chopped |
| 雜香草1束 | 1 Bouquet garni |
| 杜松果5粒 | 5 pcs Juniper berries |
| 黑胡椒粒15粒 | 15 pcs Whole black pepper corn |
| 紅酒醋200毫升 | 200ml Red wine vinegar |

## 製法 *Procedures*

**1** 把所有醃料(除紅酒醋外)混合烹煮15分鐘,放涼。

**2** 拌入紅酒醋,放入雞肉,淋上一半菜油,加上蓋,放在冰箱 冷藏一晚。

**3** 取出雞肉,瀝出醃料。

**4** 煮滾醃汁,撇去面層血水。

**5** 燒熱剩下的菜油,爆炒雞肉至金黃。

**6** 在雞肉上撒上麵粉,放入餘下的材料,以中火拌炒3分鐘。

**7** 加入熱醃汁,再煮至滾。

**8** 以慢火炆1小時或直至雞肉變軟。

**9** 加入蘑菇再煮15分鐘,調味。

**10** 放置15分鐘後即可食用。

**1** Cook marinated ingredients except vinegar for 15 minutes, let cool.

**2** Stir in vinegar, add chicken, pour over half of the oil, cover and refrigerate overnight.

**3** Remove chicken and strain the marinade.

**4** Boil the marinade and skim any blood from the surface.

**5** Heat the remaining oil, sauté chicken until brown.

**6** Sprinkle chicken with flour and add the remaining ingredients, stir over medium heat for 3 minutes.

**7** Add the hot marinade and bring to the boil.

**8** Simmer for 1 hour or until tender.

**9** Add mushroom and stand for 15 minutes, season to taste.

**10** Stand for 15 minutes before serving.

*Coq au Vin*

# 紅酒燴雞

## ➔Tips

1. 以濃縮和煮熱的紅酒醬汁長時間燉煮雞塊，可以把雞塊煮得很軟稔。這是法式經典酒燜子雞coq au vin的烹調秘密。

2. 為了讓這道菜的味道更好，做好後可以放一個晚上，食用前再充份加熱。

1. Chicken pieces are tenderized by long-period simmering in concentrated and cooked red wine marinade. This is the secret of the classic French coq au vin.

2. For a better flavour, let the coq au vin cool overnight, and reheat well before serving.

## 材料 *Ingredients*

4人
**4 pax**

| | |
|---|---|
| 雞肝370克 | 370g Chicken liver |
| 無鹽牛油190克（軟身） | 190g Unsalted butter, softened |
| 白蘭地45毫升 | 45ml Brandy |
| 暖牛油清50克 | 50g Warm clarified butter |
| 鹽、鮮磨胡椒少許 | + Salt, pepper from the mill |

## 製法 *Procedures*

**1** 用⅓的牛油把雞肝炒3分鐘至變色。

**2** 離火，加入餘下的牛油和白蘭地，用食物攪碎機攪成茸。

**3** 加入適量調味料。

**4** 把雞肝醬放到塗了牛油的盅內，把表面弄平滑，塗上牛油清，放涼，然後放入冰箱冷藏。

**5** 伴以薄多士食用。

**1** Cook chicken liver in W butter for about 3 minutes until colour changed.

**2** Remove from heat and pur e in food processor with the remaining butter and brandy.

**3** Add seasonings to taste.

**4** Transfer the paste to buttered dish, smooth the surface and then cover with clarified butter, cool, then chill in refrigerator.

**5** Serve with thin toast.

102

# *Chicken Liver Pâte*

# 雞肝醬

**⇒Tips**

1. 這是個做雞肝醬的簡易方法，雞肝柔和的口味和綿軟的口感都得以保留。

2. 不要把雞肝煮過熟，否則會變硬。

1. This is an easy way to make a pâte. It helps to bring out the real taste and soft texture of chicken liver.

2. Do not over-cook the chicken liver or they will become tough.

| | |
|---|---|
| 鴨胸3塊（每塊300克） | 300g x 3 Duck breast |
| 芫荽籽16粒（壓碎） | 16 pcs Coriander seed, crushed |
| 蜂蜜20毫升 | 20ml Honey |
| 鹽、鮮磨胡椒少許 | + Salt, pepper from the mill |

# 製法 *Procedures*

**1** 去掉鴨胸皮粗糙的邊緣。

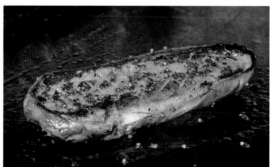

**2** 去掉鴨肉 的筋腱。

**3** 在鴨皮上劃菱形的花紋，好使脂肪可在烹調時流出。

**4** 用芫荽籽、蜂蜜、鹽和胡椒把鴨胸醃15分鐘。

**5** 不放油，鴨胸皮朝下煎4分鐘，用鏟把多餘的油脂壓出來，把鴨胸肉壓平。

**6** 把鴨肉翻過來再煎2分鐘至半熟。

**7** 從鍋裏取出，放上蓋，待用。

**8** 鴨皮朝上，刀傾斜，把鴨胸切成菱形肉片。

**1** Trim away the rough edges of the skin from duck breast.

**2** Trim away tendon from the flesh.

**3** Score diamond patterns on the skin to release fat during cooking.

**4** Marinate the breast with coriander seeds, honey, salt and pepper for 15 minutes.

**5** Place the breast with skin side down and cook for 4 minutes without oil, press with spatula to extract juice and keep breast flat.

**6** Turn the breast and cook for another 2 minutes to medium rare.

**7** Remove from pan and rest, cover.

**8** With the skin side up and knife slanted, cut thin diagonal slices.

# Pan-fried Duck Breast
# 鑊煎鴨胸

✈Tips

1. 鴨胸可以用它們自身的脂肪和汁液「乾煎」，因為鴨胸的皮下脂肪很多。

2. 鴨胸烹煮後待片刻才切片，防止肉汁流出。

1. Duck breast can be 'dry-fried' with their own fat and juice because they are so fatty.

2. Rest the duck breast for a moment before slicing to prevent the juice from running out.

鵪鶉6隻
混和油(牛油和植物油)60毫升
鹽、鮮磨胡椒少許

6 pcs Quails
60ml Mixture of butter and vegetable oil
+ Salt, pepper from the mill

### 餡料
野米和長粒米140克
雞上湯450毫升
鵝肝100克(切成小方塊)
蘑菇160克(切碎),砵酒30毫升
乾牛肝菌20克(泡在熱水中,瀝乾切碎)
乾蔥3粒(切碎),雜香草30克
(西芹,羅勒,細葉芹)(切碎)

### For the stuffing
140g Wild rice and long grain rice
450ml Chicken stock, 30ml Port
100g Foie Gras, cut into small cubes
160g Mushroom, chopped
20g Dried cep, soak in hot water,
drained and chopped,
3 pcs Shallots, chopped
30g Mixed herbs (Parsley, Basil, Chervil), chopped

### 砵酒醬汁
乾 200克(切片),牛油60克
烤鵪鶉骨架4個,黑醋50毫升
砵酒250毫升
雞上湯750毫升
新鮮百里香1枝
鹽、鮮磨胡椒少許

### For the Port Sauce
200g Shallot, sliced, 60g Butter
4 Roasted quail carcasses
50ml Balsamic vinegar
250ml Port wine, 750ml Chicken stock
1 sprig Fresh thyme
+ Salt, pepper from the mill

## 製法 Procedures

1 鵪鶉去骨,以200℃把骨架烤10分鐘
(做砵酒汁用)。

4 燒熱一些混和油,將鵪鶉四面煎黃。以200℃把鵪鶉烤15-20
分鐘,不時把熬出的肉汁澆於肉上。

5 用牛油炒香乾蔥,放鵪鶉骨架和黑醋,煮至汁液幾乎收乾。

6 拌入砵酒,煮至酒份量少一半,再加入雞湯和百里香,以慢
火炆15-20分鐘,隔出汁液,調味。

7 配上砵酒汁食用,用迷迭香和番荽裝飾。

2 填料:a)米放入上湯以慢火煮30分鐘
至軟,待用。b)用鹽和鮮磨胡椒給鵝
肝調味。c)爆炒鵝肝至外邊四面至剛
有色,加入米飯中。d)用混和油爆炒
蘑菇,放入乾蔥和香草,炒至乾蔥變
軟。e)加入米飯,調味,待用。

1 Bone the quail and roast the carcasses at 200℃ for 10
minutes (use for port sauce).

2. For stuffing: a) Simmer rice in stock for 30 minutes until
tender, set aside. b) Season the Foie Gras with salt and
pepper. c) Sauté until the cubes are sealed on all sides,
add to rice. d) Sauté mushrooms in mixed oil, add shallot
and herbs and cook until shallot is soft. e) Add rice, season to taste, let cool.

3. Season inside of quails, fill them with stuffing and truss.

4. Heat some mixed oil and brown the quails on all sides. Roast
the quails at 200℃ for 15-20 minutes, baste occasionally.

5. Sauté shallot in butter, add carcasses and vinegar, boil
until almost evaporated.

6. Stir in port and reduce by half, then add stock and thyme
and simmer for 15-20 minutes, strain and season to taste.

7. Serve the quails with port sauce and garnish with rosemary and parsley.

3 給鵪鶉內放調味料,放入餡料,紮起。

# Boned Quail, Filled with Wild Rice, Foie Gras and Mushroom

# 野米鵝肝蘑菇釀鵪鶉

➤Tips

1. 用新鮮鵪鶉，味道更佳。

2. 烘烤前才釀鵪鶉。

1. Use fresh quails to get better flavour.

2. Stuff the quail shortly before roasting.

## 材料 *Ingredients*

| | |
|---|---|
| 牛柳600克（修改好） | 600g Beef fillet, trimmed |
| 洋葱60克（切碎） | 60g Onion, finely chopped |
| 意大利番荽10克（切碎） | 10g Italian parsley, chopped |
| 酸青瓜20克（切碎） | 20g Cornichons, chopped |
| 水瓜鈕（續隨子）10克（切碎） | 10g Capers, chopped |
| Tabasco醬汁3滴 | 3 drops Tabasco sauce |
| 蛋黃6隻 | 6 pcs Egg yolks |
| 鹽、鮮磨胡椒少許 | + Salt, pepper from the mill |

## 製法 *Procedures*

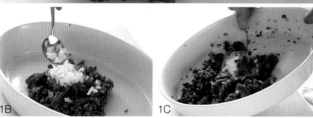

**1** Chop the beef and mix with all ingredients except egg yolk, season to taste.

**2** Shape the beef into rounds, place on individual plates and slightly hollow out the center with spoon.

**3** Slide egg yolk into the hollow, serve immediately.

**1** 牛肉切碎，與所有配料拌在一起（蛋黃除外），調味。

**2** 把牛肉弄成圓形，分開放在不同的盤子裏，中央用湯匙做一個淺洞。

**3** 把蛋黃滑到洞，立即食用。

# *Beef Steak Tartare*
# 韃靼牛肉

> Tips

1. 韃靼牛肉這道菜是在剁碎的生牛肉上放上生的蛋黃，是一道經典的法國菜。
2. 在法國，通常是與酸青瓜、水瓜鈕和 Tabasco 汁配合食用，或以芥末伴吃。
3. 只宜選用最嫩的牛柳。
4. 食用前才把牛肉剁碎。

1. Steak tartare is a dish of finely chopped raw beef topped with a raw egg yolk, it is one of the greatest French classics.
2. In France, it is usually served with cornichons, capers and Tabasco sauce, with mustard on the side.
3. Only the finest fillet steak is used.
4. Chop just before use.

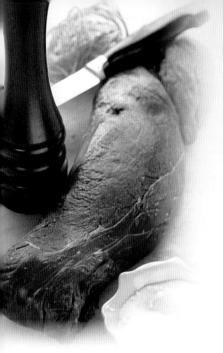

| | |
|---|---|
| 牛柳 1.5公斤（已修改） | 1.5kg Beef tenderloin, trimmed |
| 植物油 45毫升 | 45ml Vegetable oil |
| 鹽、黑胡椒碎少許 | + Salt, black pepper crushed |

## 製法 *Procedures*

**1** 牛肉修改好，用繩紮起。

**2** 用鹽和胡椒調味。

**3** 用熱油把牛柳猛火煎封表面至金黃色。

**4** 放入200℃的焗爐，焗15分鐘為生，20-22分鐘為四成熟。

**5** 焗時經常將焗盤內之熱油淋於牛柳上，保持肉質濕潤。

**1** Trim beef and tie with string.

**2** Season with salt and pepper.

**3** Sear the beef in hot oil and brown the outside.

**4** Roast beef in 200 ℃ oven for 15 minutes for rare, 20-22 minutes for medium rare.

**5** Baste frequently during roasting to keep meat moist.

# Roasting Beef Tenderloin

# 烤牛柳黐

> Tips

在焗之前先用猛火煎封修改好和紮起來的牛柳，把外層的肉煎至金黃，這樣可以保存肉汁。

Sear the trimmed and tied beef fillet in hot oil before roasting to seal and brown the outside, it can keep the meat juicy inside during cooking.

牛柳800克（焗至四成熟，放涼）　800g Beef fillet, roasted to medium rare and cooled
肝醬100克　100g Liver pâte
白蘑菇60克　60g White mushroom
牛肝菌30克　30g Cep mushroom
牛油5毫升　5ml Butter
酥皮400克　400g Puff pastry
鹽、鮮磨胡椒少許　+ Salt, pepper from the mill

雞蛋水料　*Egg wash*
蛋黃2隻　2 pcs Egg yolk
清水30毫升　30ml Water
鹽少許　+ Salt

## 製法 *Procedures*

**1** 蛋黃、水和鹽拌在一起，打至混和。

**2** 用牛油把蘑菇炒乾，調味。

3A　3B

**3** 把肝醬和蘑菇放在已涼的牛柳上。

**4** 將酥皮壓捲成4毫米厚，把整條牛柳完全包實。

**5** 牛柳卷一邊朝下，塗上蛋汁，頂部用麵條作裝飾。

**6** 牛柳卷放到焗盤上，刷上蛋汁。

**7** 用200℃焗10分鐘，然後把溫度降到180℃，再焗10分鐘至牛柳卷變金黃和酥脆。

**8** 放涼8分鐘，切片即成。

**1** Mix egg yolk, water and salt together, whisk until combined.

**2** Sauté mushroom with butter until quite dry, season to taste.

**3** Put the liver pâte and mushroom on cooled beef fillet.

**4** Roll the puff pastry to 4mm thick, then wrap the fillet to enclose it completely.

**5** Parcel side down, brush with some egg wash and decorate the top with pastry-made strips.

**6** Place the parcel on baking sheet and brush with egg wash.

**7** Bake in 200℃ oven for 10 minutes, then lower the heat to 180 ℃ and bake for another 10 minutes until golden and crisp.

**8** Let rest for 8 minutes, then cut into slices and serve.

*Baked Beef Wellington*

# 焗威靈頓牛柳

➤Tips

1. 用酥皮包住半熟的牛肉是法式的做法。

2. 烤焗前把牛肉卷放置冰箱中10分鐘，酥皮會更鬆脆。

3. 烘烤後停放片刻才切片，可保持更多肉汁。

1. Using crisp pastry to wrap medium-rare beef is a unique method in French cooking.

2. Place the rolled beef in refrigerator for 10 minutes before roasting to make the pastry more crispy.

3. Let rest before slice to make the meat more juice.

## 材料 *Ingredients*

| 小牛肉6塊（每塊120克） | 120g x 6 Veal lean meat |
| --- | --- |
| 帕爾瑪火腿18片（切薄片） | 18 pcs Parma ham, thinly sliced |
| 鼠尾草18片 | 18 pcs Sage leave |
| 牛油20毫升 | 20ml Butter |
| 橄欖油20毫升 | 20ml Olive oil |
| 馬沙拉酒60毫升 | 60ml Marsala |
| 忌廉200毫升 | 200ml Cream |
| 鼠尾草10毫升（切碎） | 10ml Sage leaves, chopped |
| 鹽、鮮磨胡椒少許 | + Salt, pepper from the mill |

## 製法 *Procedures*

**1** 把小牛肉片放在兩張保鮮紙之間，放砧板上拍打至所需形狀。

**2** 取走保鮮紙，把肉片切成三片。

3A

3B

3C

**3** 帕爾瑪火腿和鼠尾草放在肉片上，把肉卷起來，用竹籤固定。

**4** 用牛油和橄欖油煎小牛肉。

**5** 取出小牛肉，待用。在鍋加入馬沙拉酒，與鍋的汁混和，然後加入忌廉，煮至汁液黏稠適中。

**6** 放入剁碎的鼠尾草和調味料，澆在小牛肉上食用。

**1** Lightly pound the escalope on board between two plastic films.

**2** Remove films and cut the escalope crosswise in three.

**3** Place the parma ham and sage on meat, then roll meat up and secure with wooden sticks.

**4** Pan-fry veal in butter and olive oil.

**5** Remove veal, set aside. Add Marsala and mix with juice in pan, then add in cream, reduce to the right consistency.

**6** Add chopped sage and seasonings, pour over the veal and serve.

# *Saltimbocca*
# 意式煎小牛肉

✦Tips

1. Saltimbocca 在意大利語的意思是「跳進嘴」，是一道名菜。

2. 帕爾馬火腿必須薄切，才能保持真正味道。

1. Saltimbocca in Italian means  ump into mouth  it is a famous dish.

2. Parma ham must be sliced very thin to keep the real flavour.

## 材料 *Ingredients*

| | |
|---|---|
| 小牛肘1.5公斤（切成4厘米長） | 1.5kg Shin of veal, cut into 4 cm pieces |
| 麵粉30毫升 | 30ml Plain flour |
| 橄欖油45毫升 | 45ml Olive oil |
| 甘筍3個（切丁） | 3 pcs Carrots, diced |
| 洋蔥3個（切碎） | 3 pcs Onions, chopped |
| 乾身白酒180毫升 | 180ml Dry white wine |
| 濃湯底80毫升 | 80ml Brown stock |
| 番茄碎600克 | 600g Tomatoes concasse |
| 乾雜香草5毫升 | 5ml Dried mixed herbs |
| Gremolada 香料（碎檸檬皮蒜頭和番荽）5毫升 | 5ml Gremolada (chopped of lemon zest, garlic and Parsley) |
| 鹽、鮮磨胡椒少許 | + Salt, pepper from the mill |

## 製法 *Procedures*

**1** 用鹽和胡椒給小牛膝調味。

**2** 在小牛膝上均勻地撒上麵粉。

**3** 用油煎小牛肉至金黃，取出待用。

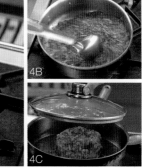

4A　4B　4C

**4** 放甘筍和洋蔥到鍋，加熱8分鐘至軟，牛膝放回鍋，加入酒、濃湯、番茄、香草和調味料，加上蓋，放進170℃的焗爐中焗2小時或直至肉變軟，期間不時翻動。

**5** 撒上 Gremolada 香料即成。

**1** Season veal with salt and pepper.

**2** Sprinkle veal with flour evenly.

**3** Cook veal in oil until golden brown, remove veal, set aside.

**4** Add carrot and onion to pan and sweat for 8 minutes, return meat to pan and then add in wine, stock, tomatoes, herbs and seasonings, cover and cook at 170 ℃ oven for 2 hours or until tender, turn occasionally.

**5** Sprinkle with Gremolada and serve.

*Osso Buco*

# 燴牛仔膝

→Tips

1. 這道著名的意大利菜源自米蘭。要做好這道菜，就要選骨多但脂肪不多的牛肘。

2. 筋腱和軟骨在烹調的過程中會被煮溶，令湯汁變稠。做這道菜我只放少量的湯水，所以烹調時要把鍋子緊緊蓋着。

3. 把肉留在湯汁中放涼，第二天再重新加熱，就會得到最好的味道和最嫩的肉質。

1. This famous Italian dish originates from Milan. To cook this dish well, select tough cuts with a good amount of bone but not too much fat.

2. Sinew and gristle will be broken down during cooking to enrich the sauce, I only use small amount of liquid in this recipe, so it should be covered tightly.

3. For the best flavour and tenderness, cool the meat in sauce and reheat the next day.

| 羊排1塊（已修改） | 1 pc Lamb rack, trimmed |
| 法式芥末30毫升 | 30ml Dijon mustard |
| 麵包糠20克 | 20g Breadcrumb |
| 迷迭香10毫升（切碎） | 10ml Rosemary, chopped |
| 鹽、鮮磨胡椒少許 | + Salt, pepper from the mill |

## 製法 *Procedures*

**1** 把肋骨之間的膜切掉，這樣羊排可以彎起來。

**2** 麵包糠與香草拌勻。

**3** 將羊排調味，以油煎封。

**4** 芥末塗在羊排上，再放上香草和麵包屑。

**5** 把羊排扭彎成皇冠的形狀。

**6** 羊排向外彎，這樣皇冠才能直立起來。

**7** 中間繫上繩子把羊排固定。

**8** 放進230℃的焗爐中焗8分鐘，然後用180℃焗10分鐘。

**9** 羊排焗至四成熟奉客。

**1** Cut membrane between ribs so the rack can be bent.

**2** Mix breadcrumbs and herbs together.

**3** Marinade the lamb and sear in oil.

**4** Spread mustard on lamb, then put herb and bread-crumbs on it.

**5** Curve the meat to form a crown shape.

**6** Bend the ribs outward so the crown can stand up straight.

**7** Tie string around the middle to hold the rib in place.

**8** Roast at 230 ℃ oven for 8 minutes, then at 180 ℃ for 10 minutes.

**9** Serve medium rare.

*Crown Roast of Lamb*

# 皇冠焗羊排

## →Tips

1. 焗皇冠羊排的法語名稱是 Couronne。

2. 這道菜名是從其外形而取的。一塊或兩塊羊排綁在一起時，看起來就像一個皇冠。這道菜是專為特別場合而設的。

1. The French name of Crown Roast is Couronne.

2. The name of this dish comes from its appearance. When one or two racks are tied together, they look like a crown. It is made especially on special occasions.

## 材料 *Ingredients*

| | |
|---|---|
| 羊腿1隻 | 1 pc Lamb leg |
| 羊肉150克（切碎） | 150g Lamb meat, finely chopped |
| 雜香草25毫升（切細） | 25ml Mixed herb, chopped |
| 蒜頭2瓣（切碎） | 2 cloves Garlic, finely chopped |
| 乾葱1粒（切碎） | 1 pc Shallot, finely chopped |
| 鮮麵包糠50克 | 50g Fresh breadcrumb |
| 雞蛋1隻（輕輕打勻） | 1 pc Egg, lightly beaten |
| 蒜頭50克（去皮） | 50g Garlic, peeled |
| 橄欖油15毫升 | 15ml Olive oil |
| 鹽、鮮磨胡椒少許 | + Salt, pepper from the mill |

## 製法 *Procedures*

**1** 羊肉、香草、切碎蒜頭、切碎乾 、麵包糠拌在一起，調味，拌入雞蛋，待用。

5

**2** 繞着肩胛骨、穿過筋腱切割，把骨頭去掉。

**3** 腿骨露出來後，繞腿骨切割，把骨頭抽出來。

**7** 在羊腿上抹上橄欖油和調味料，以220℃烤10分鐘，把溫度降至180℃再烤30分鐘。

**8** 把刀插進關節 ，切兩個很深的切口，一個橫向，一個縱向。

**9** 從切口兩邊切下肉條。

**10** 把羊腿翻過來，刀微微傾斜，切出肉片。

**5** 在羊腿上用刀做成切口，塞入蒜瓣。

**6** 用繩子把羊腿綁起來，令外形美觀，放到烤盤的烤架上。

**1** Combine meat, herbs, chopped garlic, chopped shallot, breadcrumbs and seasonings, then mix with egg and set aside.

**2** Cut around the pelvic bone and through the tendons, remove bone.

**3** Cut around leg bone when exposed, twist and pull out to remove.

**4** Spoon stuffing into the pocket, then push it in with finger.

**5** Cut deep slits in lamb leg and insert chunks of garlic.

**6** Tie the leg with string to form a neatly shape, place on a rack in roasting tin.

**7** Rub the leg with olive oil and seasonings, roast at 220℃ oven for 10 minutes, reduce to 180℃ for 30 minutes.

**8** Insert a knife into the knuckles and the joint, make two deep cuts, one vertical and one horizontal.

**9** Carve meat slices from either side of the wedges.

**10** Turn the leg over, with the knife at a shallow angle, slice the meat.

**4** 把餡料舀進羊腿肉 ，然後用手指往塞。

# *Stuffed and Roasting Leg of Lamb*

# 釀烤羊髀

## →Tips

1. 剔去骨頭的羊腿加入餡料紮起來，再放進焗爐烤。

2. 烤羊肉的時間控制：在法國，羊肉通常是半熟食用的。每450克羊肉用220℃焗10分鐘，然後用180℃焗18分鐘。內部溫度該是60℃。

3. 羊肉自焗爐取出後，放到盤子上，蓋上錫紙待10-15分鐘，這樣可以避免切肉時肉汁流出。

1. Here a tunnel-boned leg of lamb is stuffed and tied, and then roasted in oven.

2. Roasting time for lamb: In France, lamb is usually served medium rare. Roast at 220℃ for 10 minutes, then at 180℃ for 18 minutes per 450g. Internal temperature should be 60℃.

3. After removing the lamb from oven, put on tray and rest, cover with foil for 10-15 minutes, this step can avoid the meat juice from running out when carving.

| | |
|---|---|
| 豬排2公斤（去骨‧捲起） | 2kg Pork loin, boned and rolled |
| 橄欖油30毫升 | 30ml Olive oil |
| 牛奶1.5公升 | 1.5 L Milk |
| 蒜頭5瓣（壓碎） | 5 cloves Garlic, crushed |
| 鼠尾草30毫升 | 30ml Sage leave |
| 檸檬（磨碎的皮和汁）2個 | 2 pcs Lemons (grated zest and juice) |
| 鹽、鮮磨胡椒少許 | + Salt, pepper from the mill |

## 製法 *Procedures*

**1** Sear rolled pork loin in hot olive oil, turn the joint constantly to makc the fat brown evenly on all sides.

**2** Add seasonings and milk, bring to the boil, add herbs and lemon, cover and braise at 180℃ oven for 2 hours.

**3** Stir the liquid and spoon it over the pork during cooking.

**4** Slice the pork, pour sauce over and serve.

**1** 以燒熱的橄欖油封煎豬排，不停翻動，煎至各面都成均勻的黃色。

**2** 加入調味料和牛奶，煮至滾，放入香草和檸檬，蓋上，放進180℃的焗爐焗2小時。

**3** 邊煮邊將汁舀起來澆到肉上。

**4** 豬肉切片，淋上湯汁，便可奉客。

# *Braising Pork in Milk*

# 牛奶燴豬排

## ➔Tips

1. 這種不一般的豬肉調煮方法是意大利的傳統方法。在長時間的燜炖過程中，牛奶和豬肉裏的脂肪混和在一起，可以做出最美味的湯汁和鮮美的肉。

2. 用叉子來固定肉，不要弄碎！

3. 湯汁看起來應有一點點凝結的感覺，所以不要攪動。

1. This unusual method of cooking pork is a tradition in Italy. During long period of simmering, the milk is intermingled with fat in pork, it will make the most delicious sauce and succulent meat.

2. Use fork to steady the meat, don't pierce it!

3. Sauce should be slightly curdled in appearance, so don't stir it.

## 材料 _Ingredients_

**6 pax**

| 蜜桃1公斤 | 1kg Peach |
| 紅酒1公升 | 1 L Red wine |
| 砵酒200毫升 | 200ml Port wine |
| 黃糖180克 | 180g Brown sugar |
| 檸檬皮1片 | 1 pc Lemon peel |
| 橙皮1片 | 1 pc Orange peel |
| 丁香1粒 | 1 pc Clove |
| 肉桂枝1條 | 1 pc Cinnamon stick |
| 粟粉少許 | + Little Cornstarch |

## 製法 _Procedures_

**1** 蜜桃飛水去皮，但不要去核。

**2** 把紅葡萄酒、砵酒、黃糖、檸檬皮、橙皮、丁香和肉桂煮滾。

**3** 以酒汁將蜜桃浸煮5分鐘或至軟，然後離火。

**4** 讓蜜桃留在湯汁中放涼，於冰箱過一夜使之入味。

**5** 從湯汁中取出蜜桃，將汁煮至只有一半的份量。

**6** 加少許粟粉使汁液濃稠。

**7** 煮滾，用密篩過濾。

**8** 放涼，把蜜桃切片即成。

**1** Blanch peaches and remove skin but not the stones.

**2** Boil red wine, port wine, brown sugar, lemon peel, orange peel, clove and cinnamon.

**3** Use wine mixture, poach peaches for about 5 minutes or until soft, remove from heat.

**4** Allow to cool in stock and marinate overnight in refrigerator.

**5** Take the peaches out of the stock and reduce to half.

**6** Thicken the sauce with little cornstarch.

**7** Boil it up and pass it through a fine sieve.

**8** Let cool, slice the peach and serve.

*Peach in Red Wine*

# 紅酒蜜桃

→Tips

用紅酒煮水果，可使水果吸收酒精的香味和顏色。這道菜浸煮整個蜜桃，是經典的法式做法。

Fruits that are poached in wine can absorb the flavour of the alcohol and its colour. This is a classic French technique for poaching a whole peach.

| 蛋白籃料 | For Meringue rest |
|---|---|
| 蛋白 100 克 | 100g Egg white |
| 糖 200 克 | 200g Sugar |
| 杏仁片 40 克（烤香） | 40g Almond flake, toasted |
| 草莓 240 克 | 240g Strawberry |
| 藍莓 120 克 | 120g Blueberry |
| 紅莓 120 克 | 120g Raspberry |
| 紅莓汁 90 毫升 | 90ml Raspberry coulis |

| 覆盤子汁 | Raspberry coulis |
|---|---|
| 急凍覆盤子汁 250 毫升 | 250ml Frozen raspberries |
| 糖霜 20 毫升 | 20ml Icing sugar |
| 橙汁 20 毫升 | 20ml Orange juice |

## 製法 Procedures

1 蛋白和糖攪打至硬身。

2 用一個圓形唧嘴的唧筒做出籃子的底部，從中間開始，以螺旋形向上打圈做出籃子，撒上杏仁片。

3 在 100℃ 的焗爐中焗 6 小時或更長時間，直至馬鈴（meringue，蛋白酥皮）變得很硬。

4 食用時，把雜莓和紅莓汁拌勻放在蛋白籃上。

### 覆盤子汁

把所有材料以攪拌機攪成茸，以密篩過濾便成。

1 Whip up egg white with sugar until very hard for piping (A,B).

2 Use a non-sticky pump, with a round nozzle, pipe the base, working from centre in a spiral, pipe around edge to create a nest, sprinkle almond on it.

3 Bake in 100 ℃ oven for 6 hours or more until meringue is hard enough.

4 When serving, arrange berries with raspberry coulis in the nest.

### Raspberry coulis

Pur e all ingredients together until smooth, pass through a sieve.

*Berries on Meringue Nest*

# 雜莓糖霜蛋白籃

⇢Tips

蛋白置室溫下約½小時才使用，效果更理想。

Allow the egg white in a covered container at room temperature for half an hour before use to get maximum volume.

| | |
|---|---|
| 忌廉250毫升 | 250ml Cream |
| 牛奶250毫升 | 250ml Milk |
| 雲呢拿香草枝 ½ 條 | ½ pc Vanilla stick |
| 糖35克 | 35g Sugar |
| 雞蛋1隻 | 1 pc Egg |
| 蛋黃1½隻 | 1½ pc Egg yolks |
| 糖30克 | 30g Sugar |

## 製法 *Procedures*

**1** 用湯匙取出雲呢拿香草籽。

**5** 把打好的雞蛋與忌廉和牛奶混和，過濾。

**2** 取一半忌廉與雲呢拿枝和籽一起煮。

**6** 將混和物倒入準備好的焗盅中，放到烤盤上，倒進熱水於烤盤中至焗盅的一半高度，在170℃的焗爐焗 30 - 40分鐘或直至固定成形，取出放涼。

**7** 在每個焗盅上灑上糖，放在熱烤架上烤2分鐘，放涼，在1小時內食用。

**3** 煮好後倒入剩餘的忌廉和牛奶中。

**4** 攪打雞蛋、蛋黃和糖，直至糖溶解。

**1** Take out vanilla seed with spoon.

**2** Boil half of the cream with vanilla stick and seeds.

**3** Return it to the remaining cream and milk.

**4** Whip up egg, egg yolk and sugar until sugar dissolved.

**5** Combine the egg mixture with cream and milk, strain it.

**6** Pour the mixture to prepared dish, place in a roasting tray and pour in hot water till half high of the dish. Uncovered, bake in 170℃ oven for 30-40 minutes or until set, let cool.

**7** Sprinkle sugar over each dish. Put on hot grill for 2 minutes, let cool; serve within 1 hour.

# *Crème Brûlée*

# 法式燉蛋

## →Tips

要把這個法式燉蛋做得漂亮、忌廉味濃和柔滑，秘訣是把它放在雙層蒸鍋 溫和地烤焗。煮水時留意，不能讓水起泡。
於濃厚忌廉上做成焦糖脆，法語中叫作 *brûlée*。

The secret of making this French custard beautifully, creamy and smooth is to bake it very gently in bain marie. Keep an eye on the water, it should not bubbled up. Rich cream are often topped with crisp caramel, which means "brûlée" in French.

&6人
**6 pax**

| | |
|---|---|
| 黑朱古力 120克 | 120g Dark chocolate |
| 忌廉 350克 | 350g Cream |
| 蛋黃 3隻 | 3 pcs Egg yolk |
| 糖 20克 | 20g Sugar |
| 黑朗姆酒 15毫升 | 15ml Dark rum |
| 法國橘味酒 10毫升 | 10ml Cointreau |

## 製法 *Procedures*

**1** 將朱古力置於清潔之碗內，並放於熱水上慢慢攪拌，使朱古力溶化。

**2** 攪打忌廉，待用。

3A

3B

**3** 攪打蛋黃、糖和少許溫水，直至糖溶化和起泡。

**4** 用手小心地把朱古力和蛋混合物混和，加入黑朗姆酒和橘味酒以增加香味。

**5** 把朱古力混合物與攪打好的忌廉拌勻。

**6** 把混合物倒進玻璃杯中，置於冰箱中1小時便可享用。

**1** Melt chocolate in a clean bowl over hot water.

**2** Whip cream and keep aside.

**3** Whip egg yolk, sugar and a little warm water until sugar dissolved and become foamy.

**4** Combine chocolate with egg mixture by hand carefully. Flavour with dark rum and Cointreau.

**5** Mix the chocolate mixture and whipped cream thoroughly.

**6** Arrange in glasses and chill in refrigerator for one hour.

*Brown Chocolate Mousse*

# 朱古力慕士

>Tips

1. 忌廉豐富的慕斯是一道令人稱奇的甜點。

2. 溶化的朱古力在加進蛋黃液之前要稍放涼一下，否則蛋黃會凝結。

3. 拂打忌廉之器皿必須清潔無油。

1. Light, creamy mousse make a stunning dessert.

2. Melted chocolate should be cooled slightly before adding in the egg yolk mixture, otherwise, the yolk may be curdled.

3. Container for whipping cream must be clean.

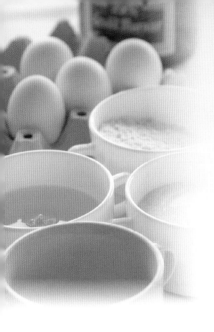

## 材料
### *Ingredients*

**6 pax**

**梳芙厘底料**

牛奶200克

糖65克

麵粉65克

牛油(室溫)65克

蛋黃5隻

**Ingredients for Soufflé base**

200g Milk

65g Sugar

65g Flour

65g Butter (at room temp.)

5 pcs Egg yolk

**梳芙厘材料**

梳芙厘底料170克

蛋黃6隻

蛋白18隻

糖160克

香料(隨意)

**Ingredients for Souffl**

170g Souffle base

6 pcs Egg yolk

18 pcs Egg white

160g Sugar

Flavour (Optional)

## 製法 *Procedures*

### 梳芙厘底製法

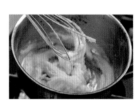

**1** 軟牛油與麵粉混和,待用。

**2** 煮滾牛奶和糖。

**3** 麵粉漿放進煮滾的牛奶中,攪打至變得黏稠。

**4** 熄火,稍微放涼,加入雞蛋黃,拌勻。

### 梳芙厘製法

**1** 在梳芙厘盆子掃上牛油,撒上糖。

**2** 將梳芙厘底料、蛋黃與香料混和,待用。

**3** 攪打蛋白和糖直至蛋白呈柔和硬度。

4A

4B

**4** 用手把梳芙厘底料與蛋白混合物拌勻,倒進梳芙厘盆子。

**5** 放進200℃的焗爐焗15分鐘。

### Procedures for Souffl base

**1** Mix soft butter with flour, set aside.

**2** Boil milk and sugar.

**3** Put flour mixture into boiling milk and whip until sticky.

**4** Turn off heat and cool the mixture slightly, add egg yolk, mix well.

### Procedures for Souffl

**1** Brush butter in soufflé dish, dust with sugar.

**2** Mix the soufflé base, egg yolk and flavour, set aside.

**3** Whip egg white with sugar until soft peak forms.

**4** Mix the soufflé mixture with egg white by hand. Pour into the soufflé dishes.

**5** Bake in 200℃ oven for 15 minutes.

# *Hot Soufflé*
# 焗梳芙厘

→Tips

1. 像忌廉大米布甸和甜梳芙厘這類柔和、爽口、歷久不衰、令人喜愛的甜點是很容易做的,能給你帶來極大的滿足感。

2. 做小梳芙厘(而非大的)的好處是,烹煮時很容易看出它們是否已熟,而且它們不容易凹陷。

3. 可隨意選配雲呢拿、朱古力、水果味或酒味的香料。

1. Like cream rice pudding and sensuous sweet soufflés, these comforting, time-honoured favourites are easy to master and can bring you great satisfactions.

2. The secret is to bake individual souffl rather than a large one  it is easy to see when they are done, and they are not easy to collapse.

3. Optional flavour can be choice of vanilla, chocolate, fruit or wine.

## 材料 *Ingredients*

**6 pax**

| 撻底料（酥皮） | Ingredients for the base(short crust pastry) |
|---|---|
| 麵粉230克 | 230g Flour |
| 水85克 | 85g Water |
| 牛油（室溫）85克 | 85g Butter (at room temp.) |
| 鹽少許 | + Pinch of salt |

| 水果撻材料（9吋模） | Ingredients for tarte tatin (9-inch mould) |
|---|---|
| 糖70克 | 70g Sugar |
| 牛油60克（溶化） | 60g Butter, melted |
| 梨4個 | 4 pcs Pear |

## 製法 *Procedures*

**1 酥皮**：把所有配料拌好，直至成一個柔滑的麵糰，待用。

**2** 梨去皮去核，切成大塊。

**3** 把糖和已溶化的牛油混和在一起，倒進撻模裏。

**5** 把梨排放在焦糖上。

**4** 用慢火煮溶，直至混合物變成焦糖，放涼。

**6** 取出麵皮，在上面掃上蛋黃。

7A

7B

7C

**1** For the base: Mix all the ingredients until they become a smooth dough, set aside.

**2** Peel pears, take out stones and cut in big chunks.

**3** Mix sugar and melted butter together, pour into the tart mould.

**4** Cook in low heat until the mixture becomes caramel. Let it cool.

**5** Arrange the pears on top of caramel.

**6** Roll out the pastry dough and brush egg yolk on it.

**7** Cover the pastry on the pears. Shape the edge of tart with knife. Arrange the pastry nicely.

**8** Bake in an 220 ℃ oven for 15-20 minutes.

**7** 把麵皮蓋在梨上，用刀修剪撻邊，好好整理麵皮。

**8** 放進220℃的焗爐焗15-20分鐘。

# *Tarte Tatin*

# 法式水果撻

## →Tips

這個經典的法國甜點是把餅皮蓋上餡面，然後翻過來食用。

This classic French dessert is cooked under a pastry lid, then served upside-down.

| | |
|---|---|
| 粉紅西柚(葡萄柚)4個 | 4 pcs Pink grapefruit |
| 糖120克 | 120g Sugar |
| 清水80克 | 80g Water |
| 青檸汁½個 | ½ pc Lime juice |

製法 *Procedures*

**1** 取出柚子瓣。

**2** 每瓣柚子切成小片。

**3** 留住西柚汁，與糖和水一起加熱。

**4** 把果汁倒在切好的柚子上。

**5** 上蓋，在溫暖的地方置2小時。
用刀輕壓果肉。

**6** 急凍至少4小時，直至果肉變
硬，中途可用叉子把果肉搗碎
幾次，這樣沙冰才會汁濃漿多。

**7** 食用時，用湯匙把沙冰從容器
中舀出來吃。

**1** Cut the wedges out of grapefruit.

**2** Chop the fillet in small slices.

**3** Keep the juice and heat it up with sugar and water.

**4** Pour the mixture over the diced grapefruit.

**5** Keep the mixture covered in a warm place for 2 hours. Crush the fillet slightly with a fork.

**6** Freeze for at least 4 hours until firm, break the mixture up with a fork several times during freezing so the granita can be slushy in texture.

**7** To serve, remove the granita from container by scraping it with a spoon.

# Iced Coulis of Grapefruit-making Granita

# 西柚碎雪

**Tips**

在冷凍過程時，要常常用叉子搗碎混合物，令到沙冰質感更豐富。

Break the mixture up with a fork as often as possible during freezing to get even and nice texture.

## 材料 *Ingredients*

| | |
|---|---|
| 薄酥皮⅓包 | ⅓ pkt Filo pastry |
| 牛油（室溫）100克 | 100g Butter (at room temp.) |
| 糖霜100克 | 100g Icing sugar |
| 打散的雞蛋100克 | 100g Beaten egg |
| 杏仁粉110克 | 110g Almond powder |
| 糖100克 | 100g Sugar |
| 青蘋果5個（去皮切丁） | 5 pcs Green apples, peeled and diced |
| ½個 檸檬（榨汁） | ½ pc Lemon juice |
| 牛油（室溫）50克 | 50g Butter (at room temp.) |
| 牛油50克（塗抹酥皮用） | 50g Butter for brushing the pastry |
| 雲呢拿香草雪糕少許（伴吃） | + Vanilla ice cream |
| 肉桂粉少許 | + Some ground cinnamon powder |

## 製法 *Procedures*

**1** 用檸檬汁和水浸蘋果丁。

**2** 攪打牛油至顏色稍變，然後與糖霜、雞蛋、杏仁粉拌在一起，待用。

**4** 把薄酥皮切成20厘米×5厘米，塗上牛油，用三張薄酥皮做一個盒子（見圖），放入一些杏仁醬和蘋果丁，包起，放在掃了油的盤子上，在180℃的焗爐中焗至金黃色。

**3** 把糖熬至成淺棕色，放入隔去水份的蘋果，煮2分鐘，加入肉桂粉拌勻，熄火，加入軟牛油拌勻，待用。

**1** Cover the apple dices with lemon juice and water.

**2** Whip butter until the color slightly change, mix with icing sugar, egg, almond powder, set aside.

**3** Cook sugar until it turns light brown, add the drained apple, cook for 2 minutes, add cinnamon, mix well, turn off the heat and mix in soft butter, set aside.

**4** Cut the filo pasty to 20cm x 5cm, brush with butter, three sliced pastries for one croustade, place some almond paste and apple dices, fold it like the picture shown above, place on an oiled tray and bake in 180℃ oven until golden.

# *Apple Croustade*
# 脆米紙焗釀蘋果

→Tips

配合雲呢拿香草雪糕食用最棒。

This dish is best served with vanilla ice-cream.

| | |
|---|---|
| 忌廉芝士 500 克 | 500g Cream cheese |
| 糖 70 克 | 70g Sugar |
| 蛋黃 2 隻 | 2 pcs Egg yolk |
| 牛奶 100 克 | 100g Milk |
| 忌廉 400 克 | 400g Cream |
| 魚膠片 6 片 | 6 pcs Gelatine sheet |
| 鹽少許 | + Salt |

**餅底料**

**Ingredients for the base**

| | |
|---|---|
| 消化餅乾 300 克（壓碎） | 300g Digestive biscuit, finely crushed |
| 牛油 70 克（溶化） | 70g Butter, melted |
| 糖 20 克 | 20g Sugar |

製法 *Procedures*

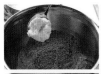

**1** 餅底：牛油溶液、糖和餅乾拌勻，壓進蛋糕模，待用。

**2** 打起忌廉，待用。

**3** 魚膠片放水中浸軟。

**4** 忌廉芝士、糖和鹽拌在一起。

**5** 倒進蛋黃和牛奶，拌勻。

**6** 把魚膠溶化，加入忌廉芝士混合物，拌勻。

**7** 再放入攪打過的忌廉，拌勻。

**8** 全部物料倒入蛋糕模中，冷藏至凝固。

**1** For the base: Mix biscuit with melted butter and sugar, press into cake mould until use.

**2** Whip up the cream and set aside.

**3** Soak gelatine sheets in water.

**4** Mix cream cheese, sugar and salt together.

**5** Pour egg yolk and milk, mix well.

**6** Melt the gelatine and add into the mixture, mix well.

**7** Fold the whipped cream into the cheese mixture and mix well.

**8** Pour all into the cake mould and refrigerate until set.

# *Chilled Cream Cheese Cake*
# 凍忌廉芝士餅

➜Tips

把忌廉芝士和攪打過的忌廉混進魚膠會為蛋糕帶來慕士般的質感，十分結實，使之易於切成片。這種芝士蛋糕也稱為雪藏芝士蛋糕，比烤焗的芝士蛋糕更清淡。

Cream cheese and whipped cream are set with gelatine to make a mousse-like texture that is firm enough to slice. This type of cheese cake is also called refrigerator cheese cake, which is lighter than the baked type.

1. **恰當的嚼勁**：將食物(一般是麵或蔬菜)煮得恰當，質感不會太軟或太硬。意大利語謂"適合入口享用"。
   **AL DENTE:** A food, usually pasta or vegetables, cooked until just firm to the bite. In Italian, it means "to the tooth".

2. **飛水 / 汆水**：將食物浸燙於沸水片刻，取出置於冰水中浸冷的過程，以停止繼續烹煮或鬆弛外皮。目的是保持色澤和去除苦澀味道，諸如番茄，桃或杏仁等。
   **BLANCH:** To immerse in boiling water, usually quick cooling in cold water, to stop loosen skins, set colour and remove bitterness of foods such as tomato, peach or almond.

3. **攪拌 / 攪碎**：用手或用小型電器(攪拌機、混合器或萬用攪拌機等)把兩種或多種材料完全混合的狀況。
   **BLEND:** To thoroughly combine two or more ingredients by hand or using an appliance such as blender, mixer or food processor.

4. **牛油**：泛指用無鹽牛油烹煮或烘烤的常用牛油，它可被橄欖油取代。
   **BUTTER:** For cooking and baking use always non-salted butter. It can be substituted by olive oil in certain dishes.

5. **焦糖化現象**：把砂糖用慢火炒至溶解，待糖色轉至金黃和發出特別香味的糖化現象。這個術語也可應用於油炒洋葱或大蒜的現象。
   **CARAMELIZE:** To stir sugar over low heat until it melts and develops a golden-brown colour and distinctive flavour. This term is often applied to onion and leek that are sautéed in fat.

6. **澄清或淨化過程**：將高湯中之雜質去除，可用撇清或加入雞蛋白於液體中讓雜質沉澱，過濾成清湯汁。牛油用慢火加熱，令到油脂與奶粉質分開，變得清徹的過程。
   **CLARIFY:** To make a liquid clear, often achieved by skimming or by adding egg white to the liquid, then strain. Also applies to the process of slowly heating butter and removing the milk solid.

7. **茸汁**：泛指已過隔篩之茸汁，一般會與番茄或含甜味的鮮果和檸檬汁混合而成。
   **COULIS:** A sieved purée or sauce, often made with tomato or fruits combined with a sweetener and some lemon juice.

8. **拂打材料至浮軟狀態**：使材料軟滑或把一至多種材料混合攪打成輕軟幼滑(浮泡)狀態。
   **CREAM:** To soften or combine one or more ingredients until the mixture is smooth and fluffy.

9. **肉或魚的薄片**：用薄切(拍打)方法處理，例如將小牛肉或魚拍成薄片狀。
   **ESCALOPE:** Thin slice of meat, such as veal or fish.

10. **主菜**：美國用語為主菜或正餐的一道菜。
    **ENTRÉE:** The main course or main dish of a meal.

11. **把食物表面塗上光澤**：可用糖水、糖霜或啫喱蓋面，或是在烹煮時或烹煮後加點油，潤澤食物。也可指為已煮濃的肉汁或魚湯。
    **GLAZE:** To coat with syrup, thin icing or jelly, either during cooking or after food is cooked. Also refers to well-reduced meat or fish stock.

12. **烘焗**：在菜式上撒上芝士碎和灑上牛油或麵包糠，置面火焗爐烘面或焗脆表面。
    **GRATIN:** A dish is topped with grated cheese and dotted with butter, or breadcrumbs, grilled or baked until crisp.

13. **撒於燴牛膝肉面上的意式雜香料**：把檸檬皮碎、蒜頭和芫荽切碎，加在意大利的燴牛膝 osso bucco 或意式燴菜上的雜香料碎。
    **GREMOLADA:** Usually made with finely chopped lemon zest, garlic and parsley, add at the end of cooking to osso bucco or other Italian stews.

14. **切幼絲**：將食物(一般是蔬菜類)切成薄幼如火柴枝似的條狀。
    **JULIENNE:** Food, usually vegetable, cut into thin, matchstick-like strips.

15. **慕士 / 泡沫狀食物**：這是一道用蛋白和忌廉混合魚、海鮮做成的泡沫狀菜式；也有用朱古力或水果做成的輕軟、美味甜品。
    **MOUSSE:** A light, delicate dish made with egg whites and cream, usually combined with fish or seafood, or with chocolate or fruit for a dessert.

16. **茸狀食物**：泛指食物轉變成液體或幼滑如醬似的物質。一般會用攪拌機攪碎。
    **PURÉE:** To convert food to a liquid or smooth, pastelike consistency. Normally using a blender.

17. **醬汁變濃 / 煎低液體**：揭起鍋蓋把食物烹煮至水份因蒸發減少的過程，可使汁液變濃而香味集中。
    **REDUCE:** To rapidly boil down liquid in an uncovered pan. This evaporates the liquid and condense the flavour.

18. **慢火烹調**：用慢火(小火)把食物放在液體中烹煮，它的液體溫度不會超過沸點，只維持鍋底或鍋邊浮出微量泡沫的狀態。
    **SIMMER:** To cook in liquid just below boiling point so that tiny bubbles form on bottom or side of pan.

19. **撇清**：除去液體表面的脂肪或雜質。
    **SKIM:** To remove fat or scum from liquid.

20. **煮軟蔬菜**：用少量油或清水稍煮蔬菜，保持軟身而不變色的狀況。
    **SWEAT:** To gently cook vegetable in fat or water until soft but not brown.

21. **圓鼓型模具 / 凍糕**：用小模具把吉士、蔬菜和飯混合而成的一度菜式。
    **TIMBALE:** A small mould is commonly used to shape custard, vegetable and rice mixture.

22. **拂打**：用手拂或攪拌器攪入空氣於混合物，造成輕軟狀和增加體積的過程。
    **WHIP:** To beat rapidly with a whisk or mixer to incorporate air into a mixture to lighten and increase volume.

23. **磨皮**：把酸性水果的有色外皮部份磨出，用作增味劑。
    **ZEST:** The coloured outer peel of citrus fruit, used to add flavour.

## 一 技術用語表 Technical Terms Table

| 中文 | 英文 | 法文 | 中文 | 英文 | 法文 |
|---|---|---|---|---|---|
| 菠蘿 | Pineapple | Ananas | 雞蛋 | Egg | Oeuf |
| 蘆 | Asparagus | Asperge | 桃 | Peach | Pêche |
| 龍蝦湯 | Lobster Bisque | Bisque de homard | 海鮮 | Seafood | Poisson |
| 法國海鮮湯 | French Seafood Soup | Bouillabaise | 蘋果 | Apple | Pomme |
| 焦糖 | Caramelize | Caraméliser | 雞 | Chicken | Poulet |
| 鴨 | Duck | Canard | 香腸 | Sausage | Saucisson |
| 甘筍 | Carrot | Carotte | 煙三文魚(鮭魚) | Smoked Salmon | Saumon fumé |
| 菇 | Mushroom | Champignon | 雪葩 | Sherbet | Sorbet |
| 潷酒 | Deglaze | Déglacer | 洋蔥湯 | Onion Soup | Soupeà l'oignon |
| 撇 | Skim | Écumer | 黑松露 | Truffle | Truffé |
| 田螺 | Snail | Escargot | 鱒魚 | Trout | Truite |
| 牛柳 | Beef Tenderloin | Filet de boeuf | 肉 | Meat | Viande |
| 鵝肝 | Goose Liver | Foie gras d'oie | 蔬菜 | Vegetable | Légumes |
| 莓 | Berry | Fraise | 甜品 | Dessert | Dessert |
| 芝士 | Cheese | Fromage | 牛仔肉 | Veal | Veau |
| 蛋糕 | Cake | Gêteau | 羊仔肉/羔羊肉 | Mutton | Mouton |
| 雪糕 / 冰淇淋 | Ice-cream | Glace | 豬排/豬扒 | Pork | Porc |
| 蠔 | Oyster | Huitre | 牛油/黃油 | Butter | Beurre |
| 火腿 | Ham | Jambon | 忌廉 / 鮮奶油 | Cream | Creme |

# 換算表 Convention Table

## 液體和固體的單位 Fluid and Solid Measurements

1公升 = 1000毫升或100分升或10厘升
1 litre (l) is divided to: 1000 milliliter (ml) or
100 centiliter (cl) or 10 deciliter (dl)

1公斤 = 1000克
1 kilogram (kg) is divided to: 1000 grams (g)

1升水重1公斤
1 litre water weighs 1 kilogram

## 英制與公制的對換 Conversion Imperial to Metric:

### 液體 Fluid

1茶匙 = 5毫升
1 teaspoon (tsp) = 5ml

1湯匙 = 15毫升
1 table spoon (tbsp) = 15ml

1液安士 = 28毫升
1 fluid ounce (floz) = 28ml

1品脫(英) = 570毫升
1 pint (GB) = 570ml

1夸脫(英) = 2品脫
1 quart (GB) = 2pt

1杯(美) = 235毫升
1 cup (US) = 235ml

1品脫(美) = 475毫升
1 pint (US) = 475ml

1夸脫(美) = 950毫升
1 quart (US) = 950ml

### 固體 Solid

1茶匙 = 5克
1 teaspoon (tsp) = 5g

1湯匙 = 15克
1 table spoon (tbsp) = 15g

1安士 = 28克
1 ounce (oz) = 28g

1磅 = 454克
1 pound (lb) = 454g

1公斤 = 2.2磅
1 kilogram (kg) = 2.2lbs

### 溫度表 Temperature Chart

| 小火 | Low Heat | 300 | 150℃ |
|---|---|---|---|
| 中火 | Moderate Heat | 350 | 180℃ |
| 大火 | High Heat | 400 | 200℃ |
| 猛火 | Very High Heat | 450 | 230℃ |

**跟大廚做西菜** **Cook Like a Chef: Western Gourmet Cuisine**

| | |
|---|---|
| 編著者 | Author |
| 鄭雅正 | Julian Cheng |
| 編輯 | Editor |
| 郭麗眉 | Cecilia Kwok |
| 攝影 | Photographer |
| 幸浩生 | Johnny Han |
| 封面設計 | Cover Designer |
| 王妙玲 | ML Wong |
| 版面設計 | Designer |
| 黎品先 | Pinxian Lai |

出版者　Publisher

萬里機構・飲食天地出版社　Food Paradise Publishing Co., an imprint of Wan Li Book Co Ltd.
香港鰂魚涌英皇道1065號東達中心1305室　Room 1305, Eastern Centre, 1065 King's Road, Quarry Bay, Hong Kong
電話　Tel: 2564 7511
傳真　Fax: 2565 5539
網址　Web Site: http://www.wanlibk.com

發行者　Distributor

香港聯合書刊物流有限公司　SUP Publishing Logistics (HK) Ltd.
香港新界大埔汀麗路36號中華商務印刷大廈3字樓　3/F., C & C Building, 36 Ting Lai Road, Tai Po, N. T., Hong Kong
電話　Tel: 2150 2100
傳真　Fax: 2407 3062
電郵　E-mail: info@suplogistics.com.hk

承印者　Printer

中華商務彩色印刷有限公司　C & C Offset Printing Co., Ltd.

出版日期　Publishing Date
二〇一三年四月第四次印刷　Fourth Print in April 2013

版權所有・不准翻印　All rights reserved. Copyright © 2006, 2010 Wan Li Book Co. Ltd.
ISBN 978-962-14-4426-4

本書初版《西廚教室》於2006年出版

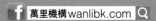

# 鳴謝
## *Acknowledgements*

- 赤柱露台餐廳 Natural Gourmet Restaurant
- Café Dido